Praise for *THE QUICKENING: CREATION AND COMMUNITY AT THE ENDS OF THE EARTH*

"Elizabeth Rush's *The Quickening* is one part memoir, one part reporting from the edge—think Elizabeth Kolbert's *The Sixth Extinction*—a book that feels as though it was written from the brink. In this case the extreme scenario is literal: Rush, a journalist, joins a crew of scientists aboard a ship headed for a glacier in Antarctica that is, like much of the poles, rapidly disappearing. The book brings the environmental crisis into a personal sphere, asking what it means to have a child in the face of such catastrophic change. [. . .] Rush writes with clarity and precision, giving a visceral sense of everything from the gear required to traverse an arctic landscape to the interior landscape of a woman facing change both global and immediate." **—*Vogue*, "Most Anticipated Books of 2023"**

"Ranging from glaciers to what grows within, this journey to Antarctica is like none you've read before—delightful and devastating, profound and grounded, but most of all shimmering with life. *The Quickening* is a mesmerizing ode to the power of melting ice and the necessity of creation amid world-altering change. I cried and laughed from cover to cover." **—Bathsheba Demuth, author of *Floating Coast***

"In *The Quickening*, Elizabeth Rush offers readers a symphony of voices from the people who stand at the forefront of climate investigations, woven with the singular lyrical story about a woman's embodied hope for the future. On a ship bound for the uncharted edge of the fragile Thwaites Glacier, experience an Antarctic voyage you've never heard before, about a warming world breaking apart, even as new life begins."
—Meera Subramanian, author of *A River Runs Again*

"The fascinating inside story of climate science at the edge of Antarctica. [. . .] The scientists are not the only heroes of Rush's book, which emphasizes above all the collaborative and interdependent nature of such voyages, where so much depends on the staff and crew. In addition to her own poetic voice, the author incorporates the voices of everyone on the ship, highlighting women and racial and ethnic minorities, who have been overlooked in the canon of Antarctic literature." —***Kirkus Reviews***

Praise for *RISING: DISPATCHES FROM THE NEW AMERICAN SHORE*

FINALIST FOR THE PULITZER PRIZE IN GENERAL NONFICTION
WINNER OF THE NATIONAL OUTDOOR BOOK AWARD
A *GUARDIAN*, NPR's *SCIENCE FRIDAY*, *PUBLISHERS WEEKLY*, AND *LIBRARY JOURNAL* BEST BOOK OF 2018

"The book on climate change and sea levels that was missing. Rush travels from vanishing shorelines in New England to hurting fishing communities to retracting islands and, with empathy and elegance, conveys what it means to lose a world in slow motion. Picture the working-class empathy of Studs Terkel paired with the heartbreak of a poet."—***Chicago Tribune***

"A sobering, elegant look at rising waters, climate change, and how low lying areas and the vulnerable people who live in those areas are at risk."
—Roxane Gay, author of *Hunger*

"Sea level rise is not some distant problem in a distant place. As Rush shows, it's affecting real people right now. *Rising* is a compelling piece of reporting, by turns bleak and beautiful." **—Elizabeth Kolbert, author of *Under a White Sky***

"These are the stories we need to hear in order to survive and live more consciously with a sharp-edged determination to face our future with empathy and resolve. *Rising* illustrates how climate change is a relentless truth and how real people in real places know it by name, storm by flood by fire." **—Terry Tempest Williams, author of *Erosion***

"With tasteful and dynamic didactic language, [Rush] informs the layperson about the imminent threat of climate change while grounding the massive scope of the problem on heartfelt human and interspecies connection." —***Los Angeles Review of Books***

THE QUICKENING

ALSO BY ELIZABETH RUSH

Rising
Still Lifes from a Vanishing City

THE QUICKENING

CREATION AND COMMUNITY AT THE ENDS OF THE EARTH

ELIZABETH RUSH

MILKWEED EDITIONS

Published 2023 by Milkweed Editions
Printed in Canada
Cover design by Mary Austin Speaker
Cover photo by Elizabeth Rush
23 24 25 26 27 5 4 3 2 1
First Edition

Library of Congress Cataloging-in-Publication Data

Names: Rush, Elizabeth A., author.
Title: The quickening : creation and community at the ends of the Earth / Elizabeth Rush.
Description: First edition. | Minneapolis : Milkweed Editions, 2023. | Includes bibliographical references. | Summary: "An astonishing, vital book about Antarctica, climate change, and motherhood from the author of Rising, finalist for the Pulitzer Prize in General Nonfiction"-- Provided by publisher.
Identifiers: LCCN 2023002380 (print) | LCCN 2023002381 (ebook) | ISBN 9781571313966 (hardcover) | ISBN 9781571317421 (ebook)
Subjects: LCSH: Antarctica--Discription and travel. | Antarctica--Environmental conditions. | Nature--Effect of human beings on--Antarctica. | Explorers--Antarctica. | Climatic changes. | Women and the environment. | Motherhood.
Classification: LCC G860 .R87 2023 (print) | LCC G860 (ebook) | DDC 998.9--dc23/eng/20230505
LC record available at https://lccn.loc.gov/2023002380
LC ebook record available at https://lccn.loc.gov/2023002381

Milkweed Editions is committed to ecological stewardship. We strive to align our book production practices with this principle, and to reduce the impact of our operations in the environment. We are a member of the Green Press Initiative, a nonprofit coalition of publishers, manufacturers, and authors working to protect the world's endangered forests and conserve natural resources. *The Quickening* was printed on acid-free 100% postconsumer-waste paper by Friesens Corporation.

For my mother.

And for Nicolás,

light among all this light.

CONTENTS

"You can't count how much we owe one another. It's not countable. It doesn't even work that way. Matter of fact, it's so radical that it probably destabilizes the very social form or idea of 'one another.'"

—FRED MOTEN

CAST OF CHARACTERS

Fifty-seven people sail to Thwaites Glacier on the R/V *Nathaniel B. Palmer* in January 2019. Roughly half are crew members and Antarctic support staff; the other half are federally funded scientists and members of the media. Below is a list of those onboard, abridged to include only those who appear in this book; it is organized around work groups and includes each person's country of origin. The scientists belong to three teams. The GHC and THOR teams reconstruct past ice sheet behavior through geologic records found in rocks and sediments. Members of TARSAN observe the present-day interaction between the ocean and the ice. Together their work informs our understanding of the current and future behavior of Thwaites, an undertaking that would be impossible without the attention, creativity, and care of the *Palmer*'s crew.

CREW:

George Aukon (Latvia): Electronic Technician
Brandon Bell (United States): Captain
Barry Bjork (United States): Electronic Technician
Cindy Dean (United States): Lab Manager, later Marine Project Coordinator
Jermaine Delacruz (Philippines): Able-Bodied Sailor
Jack Gilmore (United States): Cook
Carmen Greto (United States): Marine Technician
Julian Isaacs (Jamaica): Cook
Chris Linden (United States): Network Administrator
Lindsey Loughry (United States): Marine Project Coordinator

Fernando Naraga (Philippines): Able-Bodied Sailor
Kiel Naylon (United States): Third Engineer
Josephine "Joee" Patterson (United States): Marine Technician
Richard "Rick" Wiemken (United States): Chief Mate
Luke Zeller (United States): Third Mate

GHC:
Scott Braddock (United States): PhD Student in Geology
Meghan Spoth (United States): Master's Student in Geology

THOR:
Rachel Clark (United States): PhD Student in Marine Geology
Victoria Fitzgerald (United States): PhD Student in Geology
Alastair "Ali" Graham (Great Britain): Marine Geophysicist
Kelly Hogan (Great Britain): Marine Geophysicist
James Kirkham (Great Britain): PhD Student in Marine Geophysics
Rob Larter (Great Britain): Chief Scientist
Rebecca "Becky" Totten (United States): Paleoclimatologist

TARSAN:
Jonas Andersson (Sweden): Submarine Technician
Mark Barham (Hong Kong): Oceanographer
Lars Boehme (Germany): Marine Ecologist and Physical Oceanographer
Guilherme "Gui" Bortolotto (Brazil): Marine Mammal Ecologist
Salar Karam (Sweden): Master's Student in Physical Oceanography
Aleksandra Mazur (Poland): Postdoctoral Student in Marine Sciences
Bastien Queste (France): Biochemist
Johan Rolandsson (Sweden): Submarine Technician
Peter Sheehan (Great Britain): Postdoctoral Researcher in Oceanography
Filip Stedt (Sweden): Master's Student in Physical Oceanography
Anna Wåhlin (Sweden): Physical Oceanographer

MEDIA:
Carolyn Beeler (United States): Broadcast Journalist
Jeff Goodell (United States): Print Journalist
Tasha Snow (United States): Media Coordinator

Thwaites Glacier (Antarctica): The most important character in this play. The one that moves all the others.

Prologue

THE AUTHOR'S MOTHER:

My due date was May 25th. That's what they told me, expect the baby around May 25th. I woke up and went to work. I worked all through my pregnancy at the Massachusetts Council on the Arts and Humanities. I remember at sign-in somebody said to me, *I thought today was your due date.* I said, *It is, but I'm not having contractions or anything. So here I am.*

All during my pregnancy, I experienced a lot of anxiety. We had just purchased an old colonial house that hadn't been maintained in fifty years. We had vines growing in the windows, squirrels living in the chimneys. I looked at your dad and said, *I don't know what we've done, but we've lost our minds.* We had just enough money to buy the house. Heating it was practically out of the picture. We closed off one room downstairs and stayed there in the evening with a fire burning, and then we'd run up to the bedroom, which became your nursery, to sleep. I remember that winter we would often see our breath inside.

So I worked on my due date, but I did leave a little bit early, in the afternoon. I used to get really tired by the end of the day. I went home and I started to roam around the neighborhood to say hi to people. At each place I went, they had something they were eating. One neighbor had baked beans and hot dogs. So I had some too. One neighbor had brownies and ice cream and hot fudge. I helped myself to a big bowl.

I went to bed at about nine, and I woke up at eleven, and I said to your dad, *I think I'm having contractions.* I called the hospital. They said, *Don't worry. If they speed up, call us back.* I made it through the night, dozing in and out. I didn't get up and do anything. About five-thirty or six, I called the hospital again. They told me it was time to come in. I said to your father, *Let's go.*

Not so fast. Okay. This is a part of your birth story. Your father, he goes in and out of his closet, putting on a variety of different shirts. Asking me, what did I think of this shirt? Was he going to be too hot in it? Was he going to be too cold? Finally, I looked at him and said, *I don't care. Get dressed. Let's go.* Then he said, *I think I'd better eat a bowl of cereal.* I said, *What?* He said, *Yeah, I think I'd better eat a bowl of cereal. I don't want to be hungry.*

At that point, I was in pain. It was doable, but I was in pain. Anyway, he eats his cereal, and then he says, *Oh, we have to take a self-portrait in the nursery.* If I'd had a knife, I think I would have stabbed him because I didn't want to do that, but I went along with it, and we have a photograph. It's in our photo book. He's behind me and I'm smiling this sickly, in-pain smile in front of your crib. Then I looked at him and I said, *Time's up. We're done.*

I was getting onto the stretcher to be taken up to delivery when my water broke. That was it: *whoosh.* It was like the ocean was taking leave of my body. I went into labor and, and . . . hm . . . I'm trying to think. I did okay in the first part. My nurse said, *Let me give you something. Do you want to take the edge off?* I said, *Yeah, that would be great.* So they gave me something and I was—I can't remember how many centimeters I was dilated. There was another point where she said, *You're getting ready to go into transition, do you want this other drug?* And I said, *No, I think I'm doing okay.* If I had known—you know, in hindsight—I would have said, *Give me as much of that drug as you possibly can.*

I went into transition, which was like being summoned to hell and you can't get out. I mean, I'd never felt pain like that. The contractions were on top of each other and there's no—you can't catch your breath. You're just in it. And, for me, the problem was it also made me sick. I would throw up and, you know, I'm in it, and I'm thinking, *Okay. That was stupid. Why did you want to submit yourself to this?* But I didn't know. I didn't know.

I love it when people say, *Oh, but you just forget about the pain after.* When someone says that to me, I look at 'em and say, *Not me. No, no, not me. I remember.*

You have to crawl through transition on your hands and knees. There's no other choice. Your body does it without you, I mean, without your controlling any of it.

Once I got through it, they said, *You can push.* What happened, consistently, was you would go down, then your forehead would hit my pelvic bone. I could only get you to that certain point. After you were born, you had this red spot on your forehead, a little hematoma, from where you got caught up. Oh, I wanted to put makeup on it, because, you know, it took almost two years for it to go away.

What they did eventually was they gave me an episiotomy. That's where they make a cut in the vagina, so that there's more area for the baby to come out. I don't know if that helped, but eventually, at 1:06 in the afternoon on May 26th, you came out, you were born. Daddy picked you up and you peed on his arm and on his shirt. I held you, yes, but I didn't breastfeed you right away. Then they sewed me up. And you were a little jaundiced, so you needed to go under a light. I mean, I was in a daze through most of it afterward.

What I do remember is I had a roommate who decided I was her servant. She called me Brenda. She had something like seven children, all C-sections, and she said, *Brenda, this time they cut me deeper and harder than they've ever cut me before. Can you get me a glass of water?* And I'm like, *Sure, why not?* I got up and got her a glass of water. I wanted to say: *I've had my vagina sliced and I'm sitting on a Kotex pad with an ice pack in it, so I'm not feeling real good, but sure, let me help you out.* Your father came in. I looked at him and said, *Get me outta here.* But we had to stay one night because they wanted you under those lights. My roommate played cartoons all night long on the TV. I thought I would kill her.

The next day we went home, and it rained and rained. It rained for an entire week, the first week of your life. We just cuddled up in your bedroom. We would crawl into that bed and make a fire in the fireplace and just lie there. And Dad would make dinner and bring it up. Eventually, it stopped raining, and we went outside.

What I remember most is looking at you in the middle of the night. You were all wrapped up, the size of a football. I just looked at you and thought, *You belong to the world, and I will be your guide and your protector.* What an honor. What an incredible honor I have been given. And here we are.

ACT ONE

Part One | *Departures*

SETTING: Punta Arenas, Chile. The third-southernmost city in the world, where, in January, the sun sets near midnight. A bunch of imperial shags—cormorants clad in feather tuxedos, with disarmingly bright blue rings around their eyes—vie for space on a pier. Even farther south, the ice covering the last continent melts. Some call this moment, and the many others that are piling up, the beginning of the end. Some suspect this is how the insurgency starts, with the rattling of glaciers.

Who knows when this all got started? When we became so tangled up in each other, in ice, in obsessing with endings already in motion and what it means to make a little life while the junk drawers overflow and the jellyfish heap up on the shore and the pollinator plants just keep blooming, even deep into October, long after the monarchs ought to be gone? What do we make of all that? What do we make amid all that? Each of us begins in our own way. And yet each of us begins the same.

The year I go to Thwaites Glacier in Antarctica is also the year I decide to try to grow a human being inside of my body. It is the year of becoming two: me and you. The year we all get onto that boat, my shipmates and I, the year we sail past 73° south to the untouched edge of Thwaites, is also just another year in which the ice lets go, a little more this time. Let's agree to call it a year—like the Year of Magical Thinking or the Year of Living Dangerously—though let's also agree that it may not coincide with anything that resembles a year on the calendar. It will not start and stop on a certain day, and there will not be 365 of anything. Instead time will flow sideways, the way floodwaters cover the lowest land first, and it will unspool quick as a metal cable lowering a scientific instrument down to the very bottom of the Amundsen Sea.

On our last night on solid earth, many of us sleep in a hotel called Dreams. Everything we do anticipates what we will soon be without. Some call the children they are leaving behind. Some call credit card companies, to set up automatic payments. And some head to the Colonial for drinks. One person runs along Route 9 to stretch her legs, while another runs to the market to purchase deodorant and a couple empanadas. I go to the steam room just above the hotel's casino. Then I go to the bar around the corner for my last pint, where I eye every person who enters, wondering if we will sail to Antarctica together. In the morning, I drink a glass of honeydew juice, followed by a glass of raspberry juice. I'm thinking: *When am I going to drink fresh juice again?* And, more importantly: *Is this my last chance to be alone?*

From my table at the breakfast buffet, I can see the *Nathaniel B. Palmer* tied to the pier. The research vessel looks like a winter slipper with the heel facing forward. The low stern flares into a wide bow with a relatively flat rake, so the boat can ride up on top of the sea ice we will soon encounter, forcing it to break. The *Palmer*'s hull appears as orange as the inside of a papaya, its superstructure egg-yolk yellow. The Ice Tower, a boxy room with windows on all sides, sits at the very top, a kind of crow's nest for cold weather. Just beneath it: the bridge, where the officer on watch will oversee ship operations every single minute of every single day for the next nine weeks or more. Later, I will stand in that room and ask Captain Brandon how much the *Palmer* weighs and he will tell me 10,752,000 pounds. Yet I wouldn't call the ship large. It's roughly the length of a football field, a distance most humans can cross in under a minute without breaking a sweat. I squint through the hotel's smudged window, sip my second cup of coffee, and realize I know nothing about whatever it is I've gotten myself into.

Nine months earlier, I received a cryptic missive from Valentine Kass, my program officer at the National Science Foundation (NSF). It read: *An interesting opportunity has come up. Call me in the morning.* A strong wind blew all night, stripping the cherry blossoms from the trees. Valentine didn't wait for my response; instead she rang first thing to tell me that she had spent the previous day in a planning meeting for a five-year program to study Antarctica's most important and least understood glacier, Thwaites.

"This year they're deploying an icebreaker to investigate. There's one berth remaining, and I recommended it be given to you," Valentine said. Then she asked what was the longest I'd ever been on a boat.

"Five days," I told her, confident.

"Do you think you're up for sixty?"

"Sure," I said, perhaps a little too quickly.

"Where you'd be going, it's incredibly isolated." Valentine paused, as if waiting for me to signal comprehension. "For instance, it's easier to send help to the space station than it is for us to get help to you, if you go. The Brits run Rothera, the nearest base, and it's a four-day steam from the project site if the sea ice cooperates."

"I understand," I said, though I didn't understand anything, not really.

I had been writing about climate change's early impacts on vulnerable coastal communities for nearly a decade. During that time, I visited with hundreds of flood survivors, many of whom had lost family members and homes. I listened to their stories so that I might learn from them—and better communicate—how to navigate this time of profound transformation. That there was considerable variability in the current sea level rise models was something I had come to accept. Would there be three feet of rise or six by century's end? No one knew, and I, like those I interviewed, had to learn to live with this uncertainty.

But then I read an article about Thwaites and became uncomfortable again. If Antarctica is going to lose a lot of ice this century, it will likely come from Thwaites. That's because the glacier rests below sea level, exposing its underside to warm-water incursions that are causing rapid melting from beneath. Thwaites alone contains over two feet of potential sea level rise, and were it to wholly disintegrate, it could destabilize the entire West Antarctic Ice Sheet, causing global sea levels to jump ten feet or more. In terms of the fate of our coastal communities, this particular glacier is the biggest wild card, the largest known unknown, the pile of coins that could tip the scales one way or another. Will Miami even exist in one hundred years? Thwaites will decide.

At least that is what many scientists think, which is also why *Rolling Stone* dubbed Thwaites the "Doomsday Glacier" a couple years back. But no one has ever before been to Thwaites's calving edge—the place where the glacier discharges ice into the sea—so many of our ideas about how it will behave are a mixture of science and speculation, out-of-date modeling married to increasing fear. The more we learn about Thwaites, the more profoundly we understand that many of our predictions about the speed of sea level rise are extremely tenuous, based primarily on physical processes that human beings have already observed. It is possible that at the cold nadir of the planet, in a place that no one has ever visited, let alone cataloged in the methodical way that science demands, one of the world's largest glaciers is stepping outside of the script we imagined for it, defying even our most detailed projections of what is to come.

The possibility both haunted and intrigued me, so much so that I applied to the NSF's Antarctic Artists and Writers program with the strange hope of seeing some of this transformation firsthand. I wanted to stand alongside that massive glacier, wanted to witness freshly formed bergs dropping down into the ocean like stones, so that I might know in my body what my mind still struggled to grasp: Antarctica's going to pieces has the power to rewrite all the maps.

GUI:

I was born at five minutes to six in the morning. It was raining. My dad said something about buying a television on the day I was born. He went to buy a television, and it's the television that I inherited when I went to study veterinary medicine in Lages [Brazil]. It was massive. Of course I took it with me—I loved it. Some friends, their dads would offer them bicycles, and they would ask for a car. I was grateful for anything my dad gave me. That television was as exactly old as me, and I was very happy for that. It even had a remote control with buttons that ran all the way up to the number thirteen. You could put the remote on the television or take it in your hands: it had a magnet, you know. My friends had flat-screens and I had this old television that took up half my room, but I didn't care.

Sometimes my mom says that they had me to entertain my brother. But I know that's just a joke. Or a way to get around the question. I think a lot about the decision to put someone in the world. The motivations of my parents are completely different, if not the total opposite of mine. They were guided by this conception of family. What is your purpose in life? To have a family. I think my mom, her dream was to be a mother and to be a wife. She was so good in what she was trying to do that she ended up sacrificing a lot.

But none of that is why my wife, M—, and I wanted to have Í—. People used to think that kids should be grateful for their lives, but I don't think that is actually true. It's all about decisions. It's not easy. If you like to think, then it's very hard to accept the challenge of contributing to the world with one more person and being responsible for helping this person become someone good. Although I have hope in the world, it is a very challenging place to live at a very challenging time. I can imagine Í— asking, *Why did you have me?* just as I asked my parents. And I don't know what to tell him. I am still thinking about the answer. It is, partially, a selfish decision. It's not just about him, it's about what M— and I wanted. He wasn't there. He didn't get to decide anything.

An hour after I pay the breakfast bill, a van drops me off at a mustard-yellow warehouse in front of the pier. Inside, the twenty-six scientists who will deploy to Thwaites cluster in little groups. Some clearly know each other from previous cruises and are busy catching up; others shift awkwardly from foot to foot, like teenagers attempting to make conversation at a school dance. It's easy to spot the old-timers from their matted polar fleece, whereas most of us newbies could have walked right out of a Patagonia catalog. There at the edge of things I pause to take it all in: the three-story stockroom with its orange metal shelves, this cluttered hangar from which most of the United States' ship-based scientific expeditions to Antarctica are launched; the forklift shuttling stacks of boxes around the periphery; the sign by the door, declaring that 112 days have passed since the most recent workplace-based accident.

A woman wearing a puffy platinum vest and neon-yellow running shoes nods at me, and so I nudge my way into the circle where she stands. Rob Larter, the bearded, blue-eyed chief scientist, is here, as is Bastien Queste, a physical oceanographer. I conducted Skype interviews with many of the scientists in the lead-up to the cruise, sure that each person's expedition had begun long before embarkment. If my shipmates were going to narrate this book with me, then I would need to document what they were doing to prepare. When I spoke with Bastien, he told me he was going to miss his home-brewed kombucha, "fresh veg," and going for bike rides alone. He sounded self-assured then, just this side of cocky, but his list of soon-to-be-longed-for items made me wonder if we might become friends. Rob, who has been to Antarctica nineteen times, said he wasn't packing anything special, was leaving that up to his longtime partner.

"How's the satellite imagery looking?" Bastien asks. His polarized sunglasses pushed back to keep his wavy chestnut hair from falling in his eyes.

"Overall, it's a pretty good sea-ice year," Rob says. "Things are clearing out nicely. Most of the landfast ice that's usually in the region is gone."

When most people picture Antarctica, they see a continent shaped like a brain with an arm sticking out from the stem, reaching toward South America. But this is only half the story. Every year, the ocean surface surrounding the Antarctic landmass freezes over in what is the single largest seasonal event on the planet. Sea ice peaks in September, covering almost twenty million square kilometers of the Southern Ocean in an impenetrable frozen film. During the polar night, the continent's size temporarily doubles.

If you were to look down upon Antarctica then, it would appear more like a pancake than a cerebrum. Come summer, this ice mass retracts and the inner seas open up, allowing the Weddell, the Ross, and the Bellingshausen to become briefly navigable.

The window for working in the Amundsen Sea is among the shortest, four to six weeks at best. "Everyone competes for this time of year. February, that's the month when you go to the hardest places to reach and take the biggest challenges," Ross Hein, one of the marine project coordinators, told me prior to the cruise. "Thwaites floated to the top because it's a national priority, and they decided to send the *Palmer* because it has a big operating platform that can accommodate the diversity of instruments they're going to need to bring along: Zodiacs, coring gear, even a small submarine."

Most scientific cruises in polar regions tend to focus on one particular aspect of the physical environment, be it sediment or sea-ice coverage, the ocean-ice interface or deep-sea currents. But because of the threat that Thwaites poses, and the extreme difficulty and cost of simply getting close to it, our mission, as the first people to ever survey its calving edge, is to bring back as much preliminary information as possible. The data gathered will be used to inform the science program for the remaining four years of the International Thwaites Glacier Collaboration, the largest project the United States and the United Kingdom have undertaken together in Antarctica in nearly a century.

"Right in front of Thwaites, it looks like there just might be a pretty significant opening," Rob says as he crosses his arms. Carolyn Beeler, a reporter covering the cruise for Public Radio International, holds out her microphone. I jot a few quick notes in a small pad, as does Jeff Goodell, a reporter for *Rolling Stone* who will also be sailing south. "We've never seen this part of the Amundsen ice-free before." Suddenly everyone is talking about what they hope to achieve given this unprecedented lack of ice. While it is unnerving how much of the Southern Ocean is navigable this year, it is also comforting, in a way, to stand alongside people who hear the latest bad news from the poles and don't need to voice their concern about the rate at which the climate is changing. Instead it is a given, the very reason for our being here.

A woman in a purple down jacket announces that it's time to pick up our ECWs, or extreme cold weather kits. She leads us among the giant spools of cable and shipping pallets stacked high with aluminum Zarges cases to the far corner of the warehouse.

"Zippers break," says the bearded guy who hands me an orange duffel bag stuffed with dozens of articles of government-issued outerwear, many of them duplicates. Where we are going, there are no stores, no Amazon deliveries, no opportunities to replace something that fails. If it breaks, we've got to mend it or hope that we brought along a suitable backup.

The oldest woman in the group leans over and whispers, "Try everything on to make sure it fits." She enters the changing room, which is really just a couple of pieces of plywood tacked together, and I quickly follow.

Inside, I pull a well-worn pair of work pants the color of pond scum from my bag. "Nothing like a pair of Carhartts to remind you that you have an ass and most men don't," I say to the women around me, squatting, trying to will an extra inch of give into the thick canvas. Tasha Snow, the media coordinator, is already halfway through her pile. When she steps into a pair of rain pants and pulls out the bib, I laugh. It appears as if two of her could fit inside.

Months earlier I asked a male glaciologist what to bring on the trip. He told me to "pack as if you were going to the moon," which struck me as both useless and boastful. Erika Blumenfeld, a photographer who had previously traveled to Antarctica as an artist-in-residence, gave real advice. "Pack twice as many tampons as you think you'll need. Bring glove liners and long underwear, because what the government will provide won't be made with a woman's body in mind. And bring baby wipes, because showers are sometimes infrequent and it can feel good to be able to take care of yourself, however small the gesture."

After about twenty minutes of exchanging one ill-fitting

piece of clothing for another, I have arrived at what I hope is the best possible set of options: one red windbreaker, one pair of too-big insulated Carhartt overalls, one pair of almost-too-small Carhartt work pants, some steel-toed XtraTufs, a green raincoat and matching bib, four pairs of gloves with varying thickness and ability to resist water, one balaclava, one neck gaiter, snow pants, and one bright red hat with earflaps and a little duck-shaped bill.

"Don't I need a winter coat?" I ask the counter attendant.

"Yes. But one with built-in flotation. They've got those onboard." He looks over the clothes I'm returning and adds, "You don't want the long underwear?"

"I brought my own," I reply, handing back my bag for portside delivery.

BEFORE I SET OUT FOR Punta Arenas, the federal government sent over a pile of paperwork forty pages thick. What they wanted from me was wide-ranging: medical histories, bloodwork, dental X-rays, and EKGs, plus my answers to a prying questionnaire investigating the frequency and type of alcohol I consumed, my general emotional status, and whether or not I had hemorrhoids. I was solicited to be part of the "walking blood bank" should an onboard medical emergency arise. On page eight, one thin line read: "Pelvic Exam." Pregnant people, I learned, are not allowed to sail south. I told myself that it made sense, that no one wants to have morning sickness and throw up on a glacier. Still, my stomach tightened.

I'd often heard that for most women, fertility plummets in their mid-to-late thirties. For a while, I thought, *Well, at least I have some time.* It had been reassuring until it wasn't. All at once, I seemed to have discovered myself sitting almost at the limit of the thing, wondering how much longer I had to act. And with the arrival of the government's rule, I watched my biological clock run forward almost one full year.

When I told my ob-gyn that I would need to wait, she said, "You'll be thirty-five then, which technically will make you a geriatric pregnancy." The gown's thin paper barely covered my discomfort. "I hate that term," she added quietly.

Then don't use it, I wanted to tell her. Instead I said, "Me too." Then I got dressed, walked past the pile of *Parenting* magazines, and swiped my credit card for the ten-dollar co-pay. Walked down the single flight of stairs and out into the light, aware that I was taking the first steps, however small, toward something I had wanted for a very long time.

Something about the quality of the sunlight that day reminded me of how, close to the start of my relationship with my husband, Felipe, I sat on a park bench in New York City's Chinatown to tell him something I had never told any boyfriend. I started crying before the words even left my mouth. "I want children," I said, then added, "and I'm not getting any younger, so if you don't, then—" I don't remember what he said in response, but I know he didn't run. And I certainly knew, later, that he wanted children as much as I did. In the years since, we spoke of making a family frequently, but it was always a sort of pleasurable abstract proposition, something longed for though not yet labored after. But now what was once nothing more than desire had turned into an official declaration of intent. It sat in computer script in the most recent entry in my medical file: *Patient plans to attempt conception at the close of the coming year.*

VICTORIA:

We lived in a trailer until I was two and a half or three. My mom hated that thing—she hated being broke all the time and the fact that my dad was always away working on the oil rigs. She said, *We have to do something, join the military or something. I don't want to be poor anymore.* My dad had been in the army before he met my mom. My mom, she went ahead and took the ASVAB, but they were trying to pawn off all the cooking jobs and whatnot on the girls. She was ready to do it. But just at the last minute, my dad decided to re-enroll. He became a paralegal in the army. It worked. I grew up a lot of places: Corpus Christi; Killeen, Texas; Frankfurt am Main in Germany; Fayetteville in North Carolina; San Antonio, Texas; and Fort Campbell in Kentucky.

Nine months after I graduated from college, I realized that my student loan debt was atrocious. I was like, *Oh crap, I gotta do something.* So I joined the army too. It was the uptick, the surge, of everyone going to Afghanistan. In June 2009, I commissioned. For almost seven years—three months short of seven years—I was a military intelligence officer. I spent a year in Afghanistan in RC North. Our whole goal was to work with the border police and the Afghan National Army, who we were co-located alongside. I worked with Swedes and Germans and Norwegians and Finns, kinda like the group we've got on the boat. So that was cool. It was the safest dangerous place.

I got out in October 2015 and told my husband, *I'm gonna go back to school to become a geologist.* I spent a whole year doing post-bacc work. I had to take Calc 1, Calc 2, Physics 1, Physics 2. I got a fellowship that gave me sixty-five hundred dollars to help with childcare because I had two kids at the time. Now I have three.

That baby, she beat me to finishing my master's thesis. She came May 12th of last year. I still had my fellowship, so I decided to keep going, to get my PhD. Becky, my advisor, is the daughter of one of my old professors at Kansas State. We're kind of made for each other. She has three boys and I have three girls. When I showed up for the first day of school, Becky sat me down and she started talking to me about Antarctica's ice shelves. I was wondering why she was going on and on about Thwaites and climate change and circumpolar deep water and then it dawned on me, and it was like

> a movie. We locked eyes, and I was like, *Are you . . . inviting me to Antarctica with you?* She was like, *Yeah.* How many chances do you get to go to Antarctica? I said, *Yes!* And then I thought, in the back of my head, *Dummy, you have a husband and kids.* She told me to sleep on it, but also to let her know because the ship was leaving on January 29th. And here I am.

TOGETHER WITH TASHA AND CAROLYN, as well as Victoria and two other female scientists, I walk down the long pier to the ship that will be our home for the next two months. We stop a few hundred yards shy of the *Palmer* to snap photos of the ropes, thick as arms, that keep it snug against the pier. I hold steady there, in the ship's shadow, on the threshold of a journey I still can't fully fathom. The *Palmer*'s harbor generators roar and hum. Trails of hot water vapor spill from the smokestack, making the sky beyond it go squiggly. Soon excitement takes over, urging me up the gangplank. The metal bounces beneath my feet and the wind whips my hair into my mouth. I walk by two techs wearing hard hats, deep in conversation; an engineer smoking a cigarette; and the two men who are here to sedate and tag female elephant seals. And then I pass through two hatches, studded with large metal levers I later learn are called *dogs*—as in, "dog that door," to seal the sea out when the swell is high.

As I step aboard the *Palmer*, I try to wipe the smile from my face. Years of fieldwork have taught me that first impressions are important and that, as a woman, I'll be better served by appearing reserved rather than friendly, serious as opposed to easygoing. Of the fifty-seven people sailing south, sixteen are women, a figure that would have been all but unthinkable a few decades ago. Over the last thirty years, as opportunities to travel to Antarctica have slowly increased and more women have made it to the ice, either as workers or tourists, so too has the number of books written by them about their experiences. And while none of these stories have the same following as those of the early explorers—Ernest Shackleton, Robert Falcon Scott, and the rest—it is this alternative canon that I spent my last few months on dry land exploring.

I read *South Pole Station*, Ashley Shelby's fabulous fictional account of how a resident climate change denier tests the weave of the southernmost human outpost on earth, and *Where'd You Go, Bernadette*, Maria Semple's novel about a misanthropic architect who runs away from her fame and family to spend a couple of weeks in Antarctica alone. In Gretchen Legler's nonfiction book, *On the Ice*, she writes about her love of Edward Wilson's watercolors and how she met her future wife while stationed at McMurdo Station. In each of these books, the last continent provides an environment outside the everyday, where these women encounter more essential versions of themselves, selves that are not expected to change diapers, wear heels, or appeal to others. When one of my writing students found out about my fellowship, he gifted me a massive thermos and suggested I read Dr. Jerri Nielsen's *Ice Bound*. His mother had listened to it on tape back when he was in grade school.

"It's about a female doctor who spends the winter at the South Pole and gets breast cancer," he said. Then his voice dropped to a whisper. "I think she had to cut off her breast herself."

"Great," I replied. "Women who go to Antarctica must remove mammary glands in order to survive. Now that's a message I can *really* get behind."

But what my student remembers is only half true. Nielsen does indeed develop breast cancer while stationed at the bottom of the planet. She does not perform a mastectomy, however, but a biopsy, and eventually administers chemotherapy drugs to herself until she can be evacuated some months later.

At one point, she writes a letter to her family about why she and her colleagues are at the South Pole in the first place. It is neither conquest nor science, she argues, that drew them to Antarctica. "'We' are why we are here. We are here for each other. . . . We come to understand and rely on each other in a way that is not of this century, not of this time," she says. What stays with me long after I finish reading are not the nitty-gritty details of Nielsen's medical crisis but the way she describes the community that forms around her during the long Antarctic night, the care and close attention they pay one another, especially those who ail most. Two decades after Nielsen wrote this book, during what is arguably one of the most divisive moments in modern history, I simultaneously hunger after and dread the prospect of forming such intimate bonds with strangers.

Before setting sail, when a friend asked what part of going to Antarctica made me the most nervous, I responded, only half jokingly, "It's not the cold that scares me. I'm worried about sharing a bedroom." A similar concern rises when I think about how radically my life might change should I become pregnant. What if, by having a child, I destroy the possibility of ever truly being alone? This scares me more than appearing at work with spit-up on my blouse or a year of sleepless nights, though I only admit as much to my closest friend, afraid that I will appear an unfit caretaker for the being I want to will into the world.

On the *Palmer*, a list of room assignments has been posted on the whiteboard. I run my finger down its length until I encounter my name: *Room 131*. When I reach the door, I turn the knob but it barely budges. Try again, bracing my shoulder against the heavy green metal. This time it swings open, revealing a small bedroom with a porthole at the far end. The bunk beds have gold polyester curtains; the floors are covered in the same hard teal plastic as the hallways; and the shower stall is fashioned from some kind of metal that has been lacquered in rust-resistant paint. My roommate, Carolyn, has tossed her suitcases in a pile by a chair that is fastened to the desk with a bungee cord. In the medicine cabinet, I discover a vial of homeopathic seasickness pills, three herbal tea bags, and a tub of Eucerin lotion. Gifts from residents past.

On my way down to the Baltic Room, I pass Victoria's advisor, Becky Totten, a paleoclimatologist studying sediment cores. She's about the same age as her protégée but more burdened with bags. She's got a backpack slung over one shoulder and a ukulele over the other. Behind her she drags a giant Coleman cooler covered in stickers that warn of what will be the box's temperature-sensitive contents: mud gathered from the bottom of the Amundsen Sea. We exchange greetings and continue on in opposite directions. The walls, the handrails, the floors all shine; every surface inside the ship is designed to resist salt, water, ice, and, apparently, seafloor sludge. Down in the hold, half a dozen people root around in a mound of suitcases. One man offers to help me lug my duffel back up the stairs to my stateroom. I decline, squatting to the floor and gathering the lumpy gray bag up in my arms.

"Really, *are* they staterooms?" Peter Sheehan, a postdoc with the TARSAN project, says, in a kind of come-to-Jesus voice that makes me drop my tough-gal act and chuckle.

"Did you bring *Mean Girls*?" I ask. When we spoke a couple of weeks earlier about how to survive two months at sea, he made some fairly specific suggestions: hand cream, the TV show *Blackadder*, pounds of chocolate, hundreds of bags of Yorkshire Tea, and the classic teen comedy.

"I brought all the things," Peter says with a knowing smile. He is nothing like the image of an Antarctic scientist most might conjure. No beard, no boasting, and refreshingly little to prove. He straightens his striped sweater, doubles up his infinity scarf, and disappears down the hall.

PETER:

I had a moment yesterday. Do you ever just stop and think, *What the f— am I doing?* That we are about to sail to the goddamn Southern Ocean? Why would anybody in their right mind go to Antarctica, which is like the great beyond? I just had one of those moments where I went, *What am I doing? I'm from Norfolk. I shouldn't be here.*

Most of my friends who are not scientists think I'm kind of mad. Antarctica is bad enough, but Thwaites—they look at me like, *Really, is there a need?* They find the fact that it's called a cruise particularly entertaining because, you know, a cruise to them means rattling around the Caribbean on the *QM2* with a cocktail in hand. On our cruise there is no alcohol, you have to share a cabin, you eat what you are given, and it's freezing bloody cold.

And the thing is, you can't get off if you don't like us. If there is someone you really hate, you can't get away from them. It's the same on all the [scientific] cruises, no matter where you go. At breakfast they're there, and they spent all night on the same boat as you. You have no choice but to make it work, otherwise you'd throw yourself off the side.

I think most people make that extra bit of effort to be amenable. You don't expect to become best friends with everybody, but you do have to find some substitute for what you left behind. I don't really have any hobbies. Like, who has hobbies these days? Lonely people have hobbies. I just hang around with my friends, and we talk shit and drink coffees. We go 'round House of Fraser, which is a chain department store on the slightly trashier and glitzier end of department stores. They have a coffee shop in the bottom, and to get there you have to go through women's wear, through all the lingerie and the expensive dresses. Afterward, we go upstairs and look at all the furniture. I have a thing: I hate corner sofas, and my friends always point out all the corner sofas to torment me. I guess I'm saying my friends are my entertainment.

It may surprise you to find that I'm a quite sociable person, so to cut myself off from that almost completely—well, I've got to find some sustenance somewhere. Here we are, on

> this boat, with this motley crew of people who most of us just met, and now we will embark on this incredibly intense journey that is its own kind of social experiment.

For a long time, I never thought much about Antarctica. Never thought about the fact that, until very recently, Antarctica was just an idea to us. Aristotle imagined the earth as a sphere, with bands of frigid weather swirling around the poles. Five hundred years later, Ptolemy reasoned that if the world were indeed round, there must be something very heavy at the base to keep it from tilting off its axis: a big piece of land, to serve as a kind of anchor. The name itself didn't arrive until the start of the sixteenth century and comes from Greek. The word's root, *arktikos*, means "of the great bear"—a reference to Ursa Major, the most prominent constellation in northern skies. Centuries before any human came to know this place, *antarktikós* was constructed as the antithesis of the familiar, a cold ballast ruled by strange stars.

The proposed continent at the earth's base first appeared on world maps in 1508, when Francesco Rosselli added it to his oval "planisphere" in Florence. By the end of that century, this land that humans had conjured up but had yet to see occupied one-fourth of the total area of earth, or so Abraham Ortelius, one prominent mapmaker, thought. He called it Terra Australis Incognita, "Unknown South Land," and drew a jagged line to show where the solid earth holding the planet in place *might* be. But as sailors discovered just how far the southern seas extended, the invented continent grew smaller and smaller. Eventually, in the middle of the eighteenth century, its image disappeared from many of the most popular projections.

The desire to pull whales and seals out of the ocean, to flense them and boil their fat into oil, is, in part, what drove men further and further from what they knew—until, at long last, someone saw what human beings had only ever before imagined.

Some say Nathaniel Brown Palmer, a seal hunter from Stonington, Connecticut, and the man after whom our icebreaker is named, spotted Antarctica's great glaciers first. Others say it was a Russian named Fabian Gottlieb von Bellingshausen. Either way, in 1820, both men returned north claiming to have cast their gaze upon that blinding, icebound land. In 1821, John Davis, another sealer, was the first person to set foot on the landmass. The first to spend an entire year there: those aboard the R/V *Belgica*, from 1898 to 1899. Roald Amundsen was the first person to reach the southernmost point on earth, in 1911. Soon thereafter, Sir Ernest Shackleton would be the first to attempt to cross the entire continent, a goal that was not achieved until nearly half a century later.

With no native population, those fortunate enough to have found their way south at the turn of the last century play an extraordinarily large role in determining what we talk about when we talk about Antarctica, both figuratively and literally. Open an encyclopedia of Antarctic place names, and you will quickly realize that the men associated with the continent's "discovery" inscribe much of its terrain. Despite Shackleton's failure, an entire ice shelf, spanning 384 kilometers, carries his name. As does a stretch of coast, a mountain range, and a fracture zone buried beneath the Southern Ocean. Beyond 70° south, a mountain, an island, a point, and a sea have all been christened *Bellingshausen*. *Palmer* has been stamped across an inlet, a bay, an archipelago, a big portion of the peninsula, a mountain, and even a rocky outcropping. There's the Amundsen–Scott South Pole Station and the Belgica Trough. This continent—larger than the United States and Mexico combined—is smothered in language that reflects but the tiniest of blips in earth's history.

One afternoon, shortly after learning that I was to deploy to Thwaites, I walked over to Brown University's Rockefeller Library to see what else I could learn from those people who had ventured south before. First impressions: scant. The number of books on Antarctica in the "history, geography, travel" section didn't total more than a hundred. Most had foreboding titles like *The Worst Journey in the World*,

Deep Freeze, and *The Loneliest Continent*. Back in my office, as I stacked and inspected the nearly two dozen tomes I had selected, I realized only two were authored by women (both of whom were white), a fact I suspected was representative of Antarctic literature writ large.

The introduction to Sara Wheeler's *Terra Incognita* confirmed my hunch. She writes, "Men had been quarrelling over Antarctica since it emerged from the southern mists, perceiving it as another trophy, a particularly meaty beast to be clubbed to death outside the cave." I laughed out loud, then set it aside to flip open the other book. When I laughed again, it was with the feeling that I should have known better. Like half of everything ever written about Antarctica, Caroline Alexander's *The Endurance* was, not surprisingly, a history of the famous failed Shackleton expedition, titled after the boat he lost in the Weddell Sea.

That fall I submerged myself in the Antarctic canon and the imperial impulses of those who had journeyed south before me. Some books I could not finish. Others left me pleasantly surprised. A few months in, I grew bored. The same half-dozen events—Amundsen's conquest of the pole, Scott's death eleven miles from One Ton Depot, Shackleton's miraculous return, Douglas Mawson shooting and eating his sled dogs—are woven into nearly every narrative account of the last continent's history. These are the origin stories that circulate through the only place on earth that has no original inhabitants, the tales that get told and retold by each successive generation. Their repetition builds up a set of expectations not just about what Antarctic narratives include and what they don't, but also about who belongs on the ice in the first place. The more I read, the more I realized I wanted little to do with this tradition.

JULIAN:

Originally, I'm from Jamaica, but I moved to Florida because I got married and my wife was living there. On the island, I was making twenty-five—sometimes even thirty—dollars an hour, and everything was nice and fine. You know, I could keep up. In the States, at my first job, I was pulling in eight dollars an hour, and I was like, *No way. I can't do it.* I quit that job and found another job. Nine dollars and fifty cents an hour. So I quit that too. Found another job. All in kitchens. My third position was ten dollars and fifty cents, and I was like, *Nope, I still cannot maintain with that.* You know what I mean? I have a family to support. I thought, *Naw. I'm ready to go back.*

My wife, at the time she was a teacher, but she used to do hair on the side. A friend of hers was getting her hair done and she said, *Why not try my company, Edison Chouest* [the offshore marine transportation group that owns the *Palmer*]*?* She started giving me the breakdown, and I was like, *Cool, I'm gonna check that out.* So I call the company, and they said, *We don't do interviews or hire over the phone. You gotta come to the office.* I was living in Miami and their headquarters were all the way in Louisiana. I got all my requirements together, my merchant mariner credentials and my Transportation Worker Identification card. My wife and I, we decided: we just gonna drive on faith.

It took us almost fifteen hours. We arrived about five in the morning. Slept in the car until the office was open. When we got there, we spoke to N— C—. And he told us, *Sorry, guys, but we're not hiring at this time.* I said, *Come on, man, we just drove from Miami.* He gave us a list of other companies in the area. We spent the whole day driving around. We did five or six interviews. They said they would call us, but I wasn't sure. So I said to my wife, *Let's go back one last time, let's see if N— C— has got anything new.* We went back to the Chouest offices, and as soon as we open the door to go inside, N— C— said, *I just got a miracle email from God.* We were hired, and I've been working in the kitchen on their boats ever since.

I'm a striver. You know, you put me in a difficult situation and I strive to do my best under those circumstances. On my first trip down here, I kept asking, *Has a Jamaican ever been to Antarctica? You know of any Jamaicans who have been to McMurdo?* Some part of me wanted to be the first one. But nobody knew if I was or not. At least there were no Jamaican flags hung up at the base. Maybe I should have brought one with me and left it behind.

JACK:

I originally decided I wanted to do this to better financially take care of my grandpa. Then he passed away on December 17th [2018]. Instead of doing the hitch I was assigned, I stayed home to bury him. I was gonna call Chouest and say, *You know what, I'm not working for you guys anymore.* But then I decided that, no, I am gonna go ahead, and I'm gonna enjoy the experience. I'm gonna get on a boat with a bunch of people I don't know. I'm gonna try to continue to mentally move forward with the actual days of the week, because when my grandma passed earlier last year, I got stuck. She passed in April, and suddenly it was September, and I—I was still back in April. I knew I couldn't let that happen again.

N— C—, the same guy who hired Julian, also hired me. He said, *We're going to put you on a boat with Julian Isaacs.* When I heard that name—Julian Isaacs—I don't know, his name makes him sound like a pretty cool guy. I expected him to have three or four restaurants, and I thought, you know, maybe he's just doing this for fun. I thought he was gonna be some rich guy.

JULIAN:

I wish.

JACK:

When I first walked in and met him, I was like, *Hell yeah, he's Jamaican*. I thought, *This about to be fun*.

THAT EVENING, FROM THE SECOND-HIGHEST spot on the *Palmer*, I watch the land recede. The sunlight bounces off the sea's platinum surface, marking a long, bright trail between the stern and the horizon. I expect the ship to ring a bell or blow a horn, or for someone to smash a bottle of champagne on the bow. Instead the thrusters turn on, a few lines are tossed, and our contact with South America is no more. The *Palmer* slides out from her parking spot and steams east, through the Strait of Magellan. I stand there, on the bridge wings, for a long time, hands clenching the metal railing, cold pulsing into my palms. Out in the Strait, a plume of mist rockets toward the sky. Followed by a second and a third. I imagine a baker slapping the surface of the water with flour on her hands, the puffs suspended, then gone in the wind.

Whales?

On the back deck, over a dozen people have gathered to watch the ship leave port. Seeing them, my stomach drops: these strangers and I are sailing toward Antarctica together now. We are all we've got. Tasha, one of the women with whom I shared a dressing room, wears a bright green cap with googly eyes and two knit nubbins sticking straight up, meant to make her look like a frog. Those from England have all donned matching maroon British Antarctic Survey windbreakers, its logo embroidered above the left breast. Lars Boehme, a German-born, Scotland-based physical oceanographer, sports hiking pants that unzip into shorts and black New Balance sneakers. Joee Patterson, a marine technician from Maine, wears her hair in a reverse French braid and her Palmer Station

T-shirt tucked neatly into her Carhartts. We've been together only a day, but I am already learning names and backstories.

I walk past Becky, who is busy FaceTiming with her husband and their three boys in Alabama, to the stern. There I discover Gui Bortolotto, a marine biologist from Brazil with a colorful tattoo of Icarus on his arm, hunkered behind the smokestack, watching the whales.

"There's one!" he says, pointing north.

"Another!" I shout.

"Oh!" Gui pivots his telephoto lens toward the latest trail of spray, his black beard somehow bright in the sun. Two whales roll, front to back, along the ship's port side, their fins momentarily slicing through our vision of the Strait. "I would guess that we have more than fifty animals around us, because we're seeing, I don't know, at least twenty different blows," he says. "And if you see twenty, then you have, really, at least double that."

Three more plumes shoot skyward, diaphanous in the early evening light.

"I think it's a very good sign," I say, half to myself.

"We're going away for an unusually long time." Gui gestures to Punta Arenas and the three ochre *morros* rising up behind it. This will be the last time for months that we see streets and cars and billboards and other people, all of which grow steadily smaller as we pull away. "Welcome to the sea," Gui adds, his voice roped through with wonder.

Eventually we head inside to consult the marine mammal reference book he brought along. Before working with seals, Gui studied humpbacks and their breeding grounds in the Abrolhos Bank in northeastern Brazil. "The amazing thing about those whales is that they spend almost half of the year in Antarctica or around the subantarctic islands, feeding," he tells me, "then they swim thousands of kilometers north, returning to the same place every year to have their babies."

"Are these humpbacks?" I ask.

Lars, Gui's boss, has joined us. He flips to an illustration that looks, as far as I can tell, a lot like the animals that surround the boat.

"Sei whales?" I amend.

"It's hard to know for certain," Lars responds. Then he says something very specific about ratios of visible body to fin size and shape. My gaze slides back and forth between the sea and the shiny pages held open by two sets of hands.

"Ever since I was a girl, I've loved whales."

Lars looks defeated. "Whenever someone includes something like that in their grad school applications, I almost always move them to the reject pile," he responds.

I let out a nervous laugh, but really I wish I could take my words back. Shove all nine into my grinning mouth and swallow, hard. In my planner, I wrote a note to myself for the first week of the cruise. *Be Guarded*, it read. A reminder that my habitual enthusiasm might serve as a liability at the start of this particular expedition, might make those onboard question whether I will relay their findings accurately and with care. Objectivity is valued above all else in the sciences, and too often it's suggested it is a trait that women and writers both lack. In this moment, I don't know what's worse: that I used the word *love* to describe my relationship to the animals my shipmate studies or that I referred to my own girlhood in doing so.

"The common name (pronounced *sigh*) derives from the Norwegian word for pollock, because the appearance of pollock off the Norwegian coast sometimes coincides with the arrival of these whales," I read aloud from the field guide, in an attempt to move past my error. Will the words to sound even-keeled. "Seis are arguably the fastest of the great whales. They are usually seen traveling alone or in small groups, although large unstable pods have been recorded in some areas."

Eventually I put the book down and walk away.

A handful of the other scientists mill about the bridge making small talk and sharing in the heightened sense of setting out. Over by the radiator, I lean my forehead against the scratched glass and stare at the still-bright sea. Eight o'clock turns into nine; nine o'clock turns into ten. Soon my foible begins to recede, replaced by the sense that everything to come will be singular, a step so far outside the ordinary that the most basic actions—eating breakfast, showering, simply shuffling down the hall—will, just by virtue of their happening on a ship bound for Thwaites, take on the glow of the exceptional.

A little later, I twist down the four flights of stairs that separate the bridge from the sleeping quarters, passing as I go framed flags from Edison Chouest Offshore and the NSF. Then a sequence of faded photos of penguins: a king, its footprints an irregular circle in the snow; a lone emperor staring across the ice; and finally, at the landing between my floor and the deck above it, a flock of Adélies flopping into the sea. When I reach my cabin, I put on my pajamas, brush my teeth, slide the medicine cabinet mirror to one side, and pop a meclizine tablet into my mouth.

"We prescribe the same medicine for morning sickness and seasickness," my doctor said when she handed me the prescription.

One day a couple of months before setting sail, I found myself on my hands and knees in a friend's bathtub. Along the mold-mottled bottom of their map-of-the-world shower curtain, the outline of Antarctica stretched cold and blue. It was the sole continent not divided up into a multicolor patchwork of countries and capitals. No stars marked major cities; there were no international border lines. As I stared at its simple outline, I thought about how we often celebrate Antarctica as one of the few places on the planet that nobody owns. And while that is true in a literal sense, my early research into its literary history suggested that—at least at the language level—it is clear to whom the continent belongs.

I'd found that most of the narratives constructed around Antarctica, this place so powerful it long held humans at arm's length, are driven by the twin engines of exploration and extraction, of heroics and imperial conquest. The continent is commonly compared to a virgin, clean and pure, and therefore an appropriate site for a very particular kind of subjugation. Her "broad white bosom" draws men toward it, her "impenetrable" interior the ultimate prize. How quickly one of the few remaining blank spots on the map got transformed from a vast, unknown expanse into just another backdrop against which well-educated white men make displays of derring-do.

As I worked my way through the Antarctic canon, and as the metaphors of sexual violence piled up, my boredom gave way to alienation and ultimately anger. For most of the two hundred years that have passed since Palmer (or was it Bellingshausen?) first saw the last continent, women weren't welcome. No woman participated directly in the so-called Heroic Age of Antarctic Exploration, nor were they deployed in any of the official government expeditions of the first half of the twentieth century, though plenty supported those who did explore by caring for the children they left behind. While people of color occasionally made it south, they almost always did so in service positions. The little we know of their experience—like that of the captains' wives who were sometimes brought along for companionship—mostly comes from the books white men penned.

When one friend heard that I would deploy to Thwaites on an icebreaker, she suggested taking personal defense classes; another wanted to know just how many other women would be with me on the boat. And a third sent me an article about a Boston University professor who had taunted and degraded his female PhD students alongside the glacier that carried his name. Isolation and gender imbalance are key factors contributing to the high occurrence of sexual violence on the ice to this day, but I'd also started to suspect that the limited metaphoric language used to describe the last continent bleeds into, shapes the way Antarctica is experienced not just by the few women working there but also by those who live thousands of miles away and encounter the ice primarily in words.

I stood up and rubbed the shampoo from my eyes. Started thinking about how the bodies we inhabit determine not only how we navigate the world but also which figurative language resonates. They shape the stories we tell and the futures we might imagine building. Journalist James Geary argues as much in *I Is an Other*, his investigation of how linguistic likeness impacts every aspect of our lives, from Wall Street investments to our understanding of illness and medicine. "Our bodies prime our metaphors," he writes, "and our metaphors prime how we think and act." It is not, he says, a coincidence that progress is associated with forward motion (as in, "the difficult negotiations are moving forward") since human feet don't work well in reverse. Were we to walk sideways, like crabs, then we might express "hope for a better future by saying [our] best days are still beside [us]."

I nudged the dial toward the ceiling, making the water run just a little hotter. Meanwhile another set of ideas mixed around in my mind. The Intergovernmental Panel on Climate Change (IPCC) had just released a special report suggesting that if we want to limit global warming to 1.5° C—the margin necessary to stave off significant species extinction and the drowning of certain island nations, among other things—we need to cut the amount of CO_2 we pump into the atmosphere in half by 2030. The limited amount of time remaining for us to make widespread societal change; the fact that we live in an era of profound geological transformation, unlike any human beings have ever before witnessed, the most significant part of which is playing out in Antarctica, a place where women weren't welcome until recently; my depleting egg supply: all of this information jumbled together as I lathered my body.

The problem with Antarctic storytelling is not so much that the explorer's vision of the ice is narrow, confined by their historical circumstances, but just how disproportionate a role it plays in shaping our ideas of what the continent might be, crushing the multitudes it contains, into a few recurring tropes. What if we were taught to see Antarctica not as a prize to be won but as an actor in its own right, an entity that shapes us just as much as we shape it? What if we were to think of it as animate instead of inert, beneficial as opposed to menacing, a harbinger of transformation rather than apocalypse? It occurred to me as I stepped out of the shower and toweled off that perhaps I ought to treat Antarctica not as a desolate outpost at the very ends of the earth but as a place where life begins.

I left my friend's bathroom that morning with a strange sense of resolve: before attempting to bring a human into the world, I would travel to its farthest corner, where survival is not easy, to bear witness to the impact our species is having on the only continent with no indigenous inhabitants, this place that theoretically belongs to all of us. There is so much about what will happen that I do not know. I do not know what the expedition will do to my desire to have a child, nor do I know what this desire will do to the ice, or to the story I will write about my experience. At the beginning of this year that is not exactly a year, I sit on the cusp of so very much. I might just get to see a glacier calve, I might return and make of my body a home for another. It is disorienting to simultaneously hold these two possibilities aloft in my mind—one grounded in disintegration, the other in creation. Humans have long projected that which they most desire and fear onto the ice, and I am no exception. The longer I dwell in this discomfort, drawn in two directions at once, the more fully committed I become to chronicling the course I will chart toward Antarctica and motherhood both.

Part Two | *Stalled*

SETTING: A little farther east in the Strait of Magellan, the ship docks at the Cabo Negro pier, its tanks taking on the hundreds of thousands of gallons of diesel needed to get to Thwaites and back. The task takes many hours, almost a whole day, on the lip of leaving.

BECKY:

When I began my journey as a scientist, I wanted to go to the place on the planet that's changing the most rapidly. And it's sort of surreal, but that's where we're going: the Amundsen Sea.

JACK:

This job is full of firsts for me. First time flying. First time being on a ship. I mean, I never even left the country before, and now I'm about to sail across the Southern Ocean. It's crazy.

ANNA:

In order to address the big questions, we must collaborate across nations and across fields and across personalities. And we do have our personalities. The remoteness of Antarctica, and the Amundsen Sea in particular, demands that if you want data from more than one year, you're not always the one who gets to go get it. Whenever I feel a pang, like, *Oh, this would be easier to do on my own*, I shove that down. Collaboration is what makes longer-term studies, the ones that actually illustrate the changing climate, possible.

BECKY:

When we wrote the proposal to come to Thwaites, we had several video chats with all the members of our team—we're a very international group—outlining the scope of what we want to do and how we might go about accomplishing it. The work will take years. This is just the start.

SALAR:

My name is Salar Karam. I'm from Stockholm, Sweden. The submarine team: that's us. Ever since I got into marine science, I wanted to sail on an icebreaker, to see Antarctica, goddamn it. It's something I've wanted for so many years. I'm feeling a mixture of nervousness and terrifying excitement.

RICK:

My name is Richard Wiemken; I'm from Honolulu. I've been going to Antarctica for five years. I do my two watches, I do my four hours of admin work—that's a good twelve hours—and then I have a couple hours of exercise time. The rest is dedicated to getting the sleep I need to run a vigilant watch.

This is a comfortable ship. I sleep well and have very pleasant dreams. When I go home, I'm happy for my time off and time with my family, but I'm also always looking forward to coming to work, which is kinda rare. The journey to Antarctica is one of the highlights of my career so far. I'm sold on it. The ice, the marine life, the solitude, just the beauty. It's mesmerizing.

MEGHAN:

Before coming here, I visited with second- and fourth-graders at the school where my mom teaches. They wanted to know how many types of penguins there are, whether there are polar bears. There are not polar bears. They wanted to know how cold it gets. They wanted to know if I'm going to go swimming. I am not. I don't really like the cold. I don't like the cold at all. But you know what's the most common question I got? *How many people have died in Antarctica?*

LUKE:

You know, everyone wants to see the ice, but for us, from an operational standpoint, we're taught if you can avoid the ice, avoid it. There's no reason to go after it or into it. You know, there's just no reason.

RICK:

You don't want to hit an iceberg. We will take great care not to hit an iceberg.

VICTORIA:

When my husband heard that I might get to go to Antarctica for fieldwork, he was like, *I'm so jealous; I want to go with you. If they told me all I had to do was cut off a pinky toe, I'd do it.*

CINDY:

The last kid had graduated from high school when I started to work at McMurdo Base. It's your adventures in life that make you able to come home and share with your children something that maybe they wouldn't get elsewhere. You have to go so that they can see that it's possible, so that they will go on to have adventures of their own, whatever they may be.

Ali Graham, a marine geophysicist from Great Britain, and I linger over cups of cooling coffee. We're sitting in the dining room on the main deck. This is where the galley is located as well as the dry and wet labs, the computer room, and the salinometer station. Though I don't know it now, this is the deck where the majority of our waking hours will be spent over the next two months; a compartment so low that when the ship starts rocking, the portholes will rhythmically dip below the water line.

On this first, early morning, I tell Ali about my eleventh-hour journey into town to buy four more notebooks, one *café con leche*, three pieces of marzipan, three pieces of dark chocolate, and five coconut macaroons to add to my already sizable stash of personal sundries. "I know I'll likely run out of many, if not all, of these things in the months ahead. But some part of me looks forward to it," I confide. "I want to know who I'll become without all the stuff I think I need."

"When I first started out, I brought lots of extra things, but then I learned it really doesn't matter. The ship is well stocked," Ali says, gesturing with his chin to the buffet line. The mess looks like a cluttered 1950s diner. There are five long tables fashioned from seafoam Formica, each surrounded by ten heavy stools screwed into the floor. At the center of every eating area sits a tray stuffed with sauces and condiments: teriyaki, sriracha, soy, A.1., Worcestershire, maple syrup, multiple kinds of mustard, Tabasco, honey, imitation bacon bits, and Italian dressing. At the moment it seems like overkill, but I bet being able to personalize meals will become disproportionately important as the days on the *Palmer* turn into months. Stacked beside the drink fountain are dozens of cans of soup and packets of ramen. The mini-freezer along the far wall holds ice cream bars, the counter above it lined with plastic containers full of dried fruit—apples, mangoes, papayas, and kiwi slices—for when the fresh options run out.

"Come to think of it, I did bring beard oil and a little comb," Ali adds, pulling his salt-and-pepper ducktail to a fine point. He's got a striped T-shirt under his fleece jacket, gangly limbs, an eager smile, and a disheveled pompadour hairdo. A charming hybrid of nerd and hipster, Ali expresses great interest in the way the world works while also cultivating his own sense of style.

On the wall above his right shoulder hangs a poster of an alligator in a tuxedo tending a pot of boiling crawdads. The caption reads: *Hot Tubbing, Louisiana Style*. It's not the only visual reminder of the Bayou State onboard. Scattered all around the ship are artifacts that celebrate the birthplace of Edison Chouest, the shrimper who turned his knowledge of the Gulf into a major shipbuilding and offshore petroleum service company—which, ironically, owns not only the *Palmer* but also the R/V *Laurence M. Gould*, the other icebreaker that the US government leases to study climate change at the poles. At the farthest end of the wall there's an outdated map of Chouest's home state, hung when the *Palmer* first set sail a quarter century ago. In it, the deltaic lands surrounding the Mississippi River's low-lying mouth appear fertile and green.

When I point out the gaps between the map and the Louisiana of the present—where so much of this green land is now underwater—Ali nods in quiet agreement. He's just accepted a job at the University of South Florida in St. Petersburg and plans to relocate his family there come summer. "I'm surprised by how many of my colleagues have or want ocean views. The school is on a barrier island, so, you know, that isn't going to be there forever."

"You have kids?" I ask.

"I have a two-year-old daughter and another one on the way," he says. "This is my eighth Antarctic cruise, but the first since R— was born, the first since I got married." *Not just one kid, two*, I think as a small wave of relief washes through me. Ali takes off his glasses and starts chewing on the temple tip.

Behind us, the partition that separates the buffet from the dining area is covered in memorabilia from cruises past. The *Palmer*'s smashed christening bottle is there, sealed in a glass case with a bunch of moldy ribbons. There's also a bronze statue of "Rocky, the Macaroni Penguin" and a cast of an ammonite, a tightly coiled shell that housed something like an octopus over fifty-five million years ago. Ali and I talk a little more about family and what it means to choose to make one now, but don't go very deep into the topic. We can't, just yet, but that too will change. "I worry that because Antarctic media is supersaturated, often with pieces that have no take-home message, when we do finally deliver some really important finding, it'll be diluted by all the other crap out there," Ali says a little later. "Like, *Look at this iceberg the shape of the Taj Mahal!*"

"Yeah," I say, then groan. "Most everything that's published about Antarctica is fluff or end-of-the-world stuff. I'm not convinced either is useful if you want to bring about change."

"I don't envy you," Ali says before he stands and heads over to the Dry Lab for his first meeting of the day. "Sometimes the science seems like the relatively straightforward part."

FERNANDO:

I started on the *Palmer* in February 1992, the year the ship was built. I call the ship my first home, and for me my second home is in the Philippines, because I spend more time here than I do in my province of Bohol. I come from an island that is about one hour from Manila in a plane. Some years I'm back there for six or seven months, but more often it's only two or three. It's not always the same, because it depends on the science schedule.

Before, we had an American cook who wasn't good at making the Pilipino food, so we brought some sardines, some noodle soup, some bagoong, something like that. They would just boil stuff, and it was up to us to make it taste good. It's really personal, the flavors—garlic, ginger, onions. Every one of us would bring special ingredients to help keep us alive over here. But now that we have Ariel [the third cook onboard, who is from the Philippines], we don't bring as much, because he makes the salty dried fish, the things we know and like.

There are a lot of Pilipino sailors everywhere. That's the only way we can make good money, money that compares with what you would make if you study to be a lawyer or jump into politics. If you are an ordinary worker, even if you work for a bank, you earn some money, but it's not enough. I chose to work over here so I can make good money for my family, for my son and daughters. So I can give them a chance to study.

My first time in Antarctica, we could not go close to the land because the ice was too much. Now we can go closer because the ice is melting. They say it is global warming, so I say maybe that is true, because I can see it. And I can feel the temperature too. It's warmer than before. In the beginning, this job was really hard for me, because I come from a tropical country and I didn't have any experience with the cold. But I can see that it's changing, see the glacier pull back behind Palmer Station. Back in Bohol, there are some old people, people who have eighty years or more, they're saying that the heavy rains that we're having, the mudslides on the Chocolate Hills—this is the first time in their lives it has happened. The world is different, I know.

Four months before departure, Felipe and I temporarily relocated to Bogotá, Colombia—my husband's motherland and our home base during his academic sabbatical. Sometimes during that fall, I felt as though I sat at the beginning of a new story. But just as often, I worried that by delaying pregnancy one year, I was putting it off permanently; that by choosing Antarctica now, I was saying no to motherhood later. Which is why I started writing my most basic desires down in Spanish. I was practicing the subjunctive as I filled my notebook—*Deseo que tengamos un bebé el próximo año*—but I was also attempting to talk with the other side of the universe, to remind it that I was still here, still ready and wanting to become a mother.

In those first months in Bogotá, I read Ursula Le Guin's short story "Sur" and entered an entirely different Antarctica. In Spanish, *sur* means south. South, as in the singular goal of the Heroic Age—a destination determined by countries from the global north, for whom such folly was affordable. South, as in the title of Shackleton's memoir of the doomed voyage that turned him into a legend. In Le Guin's deft hands, the word and the world it describes take a radical turn. Her female, Latin American narrator begins, "Although I have no intention of publishing this report, I think it would be nice if a grandchild of mine, or somebody's grandchild, happened to find it someday; so I shall keep it in the leather trunk in the attic, along with Rosita's christening dress and Juanito's silver rattle and my wedding shoes and finneskos."

I stopped reading and Googled "finnesko." *Finnesko*: from Norwegian, a soft boot of tanned reindeer skin used for cold-climate travel. The kind of thing someone might wear on a voyage to Antarctica. A relic of exploration that, in some mind-boggling turn, was appearing alongside a handful of domestic baubles, animal hide and sterling nested together in a chest on the top floor of a small suburban house in Lima. I felt the gendered boundary erected around Antarctica blur a little. It was exhilarating, and I was only on the first sentence.

The narrator, who does not have a name and therefore could be anyone, gathers with eight other women—one Peruvian, three Argentines, and four Chileans—in Punta Arenas in 1909. They've lied to their husbands, told them that they are spending the winter in Paris or, in a few more desperate cases, a Bolivian convent. In the "abominable seaport," they discuss who will be in charge should things go sideways and what route to take, eventually electing to aim southwest toward the Ross Sea. As I read, I felt a flutter of recognition. Soon I too would be in Punta, setting sail. For a blissful moment, I forgot that Le Guin's short story was just that, a story.

In "Sur" Juana, Carlota, Berta, and the rest cross the great

Southern Ocean in a little steamship they nickname *la vaca valiente* (or the valiant cow); they raise a glass of Veuve Clicquot when they see their first iceberg, and another when they finally spot land. They encounter Scott's famous Discovery Hut—inside of which tins of tea have been left open, the floor scattered with stale biscuits—a polar bachelor pad. Teresa suggests cleaning it up and using it as their camp; Zoe wants to set it on fire. They do neither. Berta and Eva build an ice cave instead and name the central chamber Buenos Aires. In that cold home, the women play banjo, carve ice sculptures, and plan their expedition to the pole. After months of preparation spent placing supply depots and physically acclimating to the cold and the wind, the narrator and five others set out, dragging heavy sledges into the great silence, hoping to reach the southernmost spot on earth by foot.

Like any other early explorer, they name—but in this case "not very seriously"—the land features they encounter. What Shackleton calls the Beardmore is briefly known as Florence Nightingale Glacier. Just south of it lies Mt. Bolívar's Big Nose. On December 22, 1909, they become the first people to reach the South Pole. The women talk about leaving "some kind of mark or monument," a flag perhaps, but decide there is no point. "Achievement is smaller than men think," Le Guin writes. "What is large is the sky, the earth, the sea, the soul." I underlined these words. Instead of snapping photos to commemorate their accomplishment, the ladies drink a cup of hot tea and turn around.

The real climax comes later, when, upon returning to the ice cave, the sledging party discovers that Teresa is pregnant. She goes into labor on the ice and screams herself "hoarse as a skua." The other members of the expedition tend to her, for many have labored themselves. After twenty long hours, little Rosa del Sur is born, and they drink their last two bottles of champagne toasting to her.

I closed my copy of *The Wide White Page,* an anthology of writers imagining Antarctica, and stared out the window. Instead of consecrating the closing of possibility—the title of "first" can only be claimed once for every accomplishment, after all—"Sur" celebrated the ice as an expansive thing, a place for new life and new stories. In this way the text felt generous, kicking the door open a little wider for the generations to follow. While little Rosa del Sur dies of scarlet fever at age five, her mother goes on to have many other children, and those children have children of their own. It was the first piece of Antarctic literature I'd encountered where it felt like the story continued to unfold long after I put it down. It might have been fiction, but it was also a start.

JOEE:

Up north, I work in the nonprofit wing of a wildlife refuge. It consists of about seventy islands strung along the coast of Maine. Our focus is on migratory birds. Maine is home to 96 percent of the arctic terns in the United States, outside of Alaska. That number kind of blew me away. Terns are really strong fliers, and some that are born in Maine migrate all the way to Antarctica.

This year my journey—so to speak—followed that of the birds. I left Maine in November, which is about a month after the terns have left North America, and then came down here just in time to intercept the other end of their migration. The waters that surround Antarctica are so rich in life during the southern summer. The terns get really fat on krill, and then they fly back to Maine to have their babies.

I started sailing on tall ships in New York City in 2004, just out of college. I really loved that work. It was a whole world I didn't know existed until I started doing it. I had heard about people getting jobs in Antarctica, but it took me about eight years to get my head wrapped around the idea of it and to get my ducks in a row to get hired. Now I run the machines that the scientists depend upon, do deck work, and drive the small boats that take them to shore. It's twelve hours on, twelve hours off, in this case for fifty days straight. And the anchor tattoo on my ring finger? My sister and I got them together in our twenties. We loved the idea of being married to the sea. Now I am married to an actual man, and the tattoo—let's just say it's a badge of honor.

It's complicated being a woman in this world. On my last cruise, we had this really cohesive group of women working as marine techs [MTs]. It was interesting to feel how much my brain relaxed because I wasn't doing all the calculations and mental maneuvering that I do when I am in a less gender-balanced group. I wasn't as busy trying to prove myself. The comfort I felt on that cruise wasn't just because there were more women than usual. I've had experiences with women who are older than me, who've been in this world for decades, who were forced to conform to a masculine way of doing things. Sometimes it's even harder to work with women coming from that place, because they feel the need to strong-arm you to establish their dominance. I'm sympathetic to it on a personal level, but on a professional level I have no time for it.

Back home, my friends are like, *You're so tough, your job is so badass.* And I'm like, *You should meet the people I work with, because they make me feel kind of soft.* The first couple of days have been pretty rough on me. I've learned the hard way that if I don't start my day with stretching and calisthenics, my body is going to hurt. Some of the MTs, when they're off-ship, they're swimming multiple miles a day just to stay in shape. I'm not that cool. I'm a freakin' buttercup by comparison.

THAT AFTERNOON, I HEAD FROM my cabin up to the third-floor conference room to attend a required safety training. Lindsey—the liaison between the ship's crew and the scientists, the *Palmer*'s one-woman personnel hub—plugs her computer into a screen while my shipmates shuffle in, most wearing exactly what they had on the night before. Rob is the exception. His T-shirt *du jour* looks like an entry in the periodic table. "Ah!" it reads, "the element of surprise."

"Can someone turn off the air-conditioning?" Victoria asks, zipping up her University of Alabama vest. Victoria is here to help Becky retrieve and analyze sediment cores. The position is a kind of apprenticeship: in return for her labor, she will gain field experience, a requisite should she eventually want to run a lab of her own.

A couple of people stand on chairs to close the AC vent, but it won't budge. Tasha puts on her frog hat and slumps down into the sofa opposite the ship's library shelves. They contain a bit of everything: marine science textbooks and romance novels, bird identification guides and ship's logs—but the majority of these books fall under the broad category of pulp, with titles like *Hunt Angel #7*, *Under a Bomber's Moon*, and *Crusader's Cross.* On the other side of the porthole, the land is low and flat and yellow. The sky beyond the fuel silos light blue. For the past twelve hours, the ship has been stationed here, filling up.

Since we've only just gotten underway, the extended pit stop feels like a setback. One that I have chosen to fill by sketching the Strait in my notebook and exploring the *Palmer.* There's a sauna just down the hall from my stateroom, along with a tiny gym. The second deck holds the hospital and most of the crew's quarters. The third floor, beyond the conference room, is where the captain and the chief scientist sleep. Above us there's a lounge reserved for the crew, furnished in mustard-yellow leather chairs and kitschy paintings of the very ship on which we now sit. The mates steer from the fifth and uppermost deck, also known as the bridge; and much to my surprise, anyone can visit at almost any time, even during heavy seas.

"Close the curtains," a man with a long gray ponytail calls out.

The conference room goes dark. The radio clipped to Lindsey's belt sends out a stream of static. She turns the volume down, flops her braid over her shoulder, and hits play. The informational video starts with triumphant music and Caterpillar snowplows. "The United States is one of thirty-three nations operating in Antarctica," the narrator announces, "and in 1991, these nations adopted the Environmental Protection Protocol." Cue images of ice shelves and penguins and humans sorting plastic bottles. "The United States Antarctic Program generates three and a half million pounds of waste per year, of which about sixty-five percent gets recycled."

"Not us," Cindy Dean, the lab manager, says, her voice twanging against the midday darkness. "We have to wait until we're in international waters, and then it all gets tossed into the incinerator. We don't burn our lab glass and we don't burn the chemically contaminated stuff, but everything else, all the trash and food scraps, that we burn."

The longer we sit in the air-conditioned room, the more I wish I had worn my wool hat to the meeting.

"It's still a heck of a lot colder back in Maine," says Meghan Spoth, a master's student in geology. "It's minus two up there, thanks to the polar vortex."

"Right," says Scott Braddock, who shares an advisor with her. They're here to hunt for ancient penguin bones and other organic materials on a string of islands northeast of Thwaites. Though they make a funny pair: he's got an earnest lumberjack vibe, while she is thin-boned and piquant. Across the long table, Peter's eyelids flutter shut. A few others join him for a nap. At first, I find their blasé attitude toward this safety briefing alarming. But then a second, perhaps more disturbing thought chases out the first: maybe these snoozing career scientists know that no informational video can prepare us for what lies ahead.

The afternoon concludes with an introduction to general safety on the ship. This seems important, so I start taking notes: *Personal floatation devices and steel-toed shoes are required when you are on the back deck. Any time you hear the hydraulics running, put on a hard hat. Keep one hand on the boat at all times, and never turn your back to the sea. If you see white spots like freckles on your cheeks, or if the tips of your fingers get waxy, go inside. That's frostbite.*

"If you come to me with frostbite, I will help you," Jennie Mowatt, one of the marine techs, says. "But I'll also ask you why you weren't taking care of yourself."

"There are three EMTs on the *Palmer*. Jennie, Lindsey, and myself," Cindy adds. "But I don't practice regularly. If you need medical attention, you'll be getting it from a person who will be recalling skills they learned in a book a long time ago." Jennie runs her thumb along the blade of the knife she carries in the ruler pocket of her work pants. Everyone not sleeping is glassy eyed. Something about the high number of preparatory meetings and simulations reminds me of summer camp, but with significantly greater stakes. The day before, we gathered in the same room for an emergency drill. The ship's feeble alarm bell rang: a signal that everyone was to go grab their Mustang survival suit. Mine, I learned, lived in a pillow-sized sack stowed above the drawer where I kept my underpants and thermals.

"If you hear a bell that's not been talked about, most likely it's a false alarm. But if it isn't a false alarm, then it's the real thing," said Lindsey. "It's also a good idea to bring a ditch bag. You know: a bottle of water, some high-energy snacks, a hat. Whatever you think you might need if we have to get into the lifeboats."

In the gloom of the conference room, I unfastened the plastic carrying case, and a bright-orange, neoprene onesie unrolled before me like a carpet. Overwhelmed by the undertaking, I sat down. Put my right foot in, then my left. Shoes and all.

"I hope I got a medium," said Rick, the six-foot-something chief mate. "I'm a medium kind of guy."

We'd only been onboard for a couple of hours, and already we were faking an exit. We were told we would repeat the drill throughout the journey, building muscle memory in the event of a real emergency.

"The Southern Ocean is so cold," Rick added, "that you will start to stiffen and freeze after five minutes if you go overboard and are not properly protected." What he didn't say—but what one of the MTs told me earlier that day—was that about ten years ago, a crew member went missing from the *Gould* in the middle of the Drake. Their body, never recovered. "If you see someone go overboard," Rick continued, "throw as many floatation devices and life rings at them as possible. It helps to keep an eye on the person who's in the sea."

"The suit's hood makes everyone hard of hearing," said Lindsey.

"What?" the man seated next to me asked.

"Now that we've had a one-hour class in getting dressed, we're ready to go," said Johan, the Swedish sub tech.

After wriggling both of my arms into the sleeves, which ended with gloves that made my hands look like lobster claws, I asked Carolyn to zip me up. She had not yet started to enter her immersion suit. Instead, she swiveled around her shotgun mic, attempting to capture the sounds a bunch of scientists make as they prepare to leave a sinking ship. Watching her, I was partly annoyed by the presence of her obtrusive recording device and partly grateful that my reporting sometimes allowed me to blend in, to seem the same as everyone else.

LINDSEY:

> I've been working on ships since I was eighteen years old. I just got hooked. I love the instant community. There's a different kind of interaction that happens between humans on a boat. My husband [Carmen Greto, a marine tech on the *Palmer*], he's a seagoing gent as well. It takes a certain kind of person to sign up to be away from home for months, who's willing to step into a completely different set of circumstances from life on land. I've always thought it would be interesting to bring a psychologist onboard, to find out what really happens at an interpersonal level when people are at sea for an extended period of time.

My day started at 7 a.m. The first thing I did was look out my window to see where we were. My understanding last night, before I went to bed, was that we were leaving the fuel pier to go to the prep pier and clear customs, but somewhere in the middle of the night things got switched around, because when I woke we were back at the Magallanes Pier. The captain said systems check showed an issue with the rudder, so we turned back. It's pretty rare that there's a problem with the ship.

The divers have been out. I don't know what they found yet. There was, however, a huge puddle in the hallway in front of the whiteboard and drawings of—you know—rudders. Which gave me the sense that someone had been in the water and was trying to give people information. My hope is that we will be able to get back on track soon, but I don't want to estimate a time, because when you do that you set expectations. I'll say this: a rocky start means a smooth sail.

Part Three | *First Passage*

SETTING: The rudder problem is solved. The ship sets sail a full two days behind schedule. Then perpetual sun on perpetual water. It takes ten more days to reach the last continent, and the author bleeds through all of them. Uses up more than half her supply of tampons. She wonders what her body is trying to tell her, its circuitry going as wild as the waves in the Drake.

My eyes snap open in the difficult light. I do not know whether it is day or night, nor where exactly we are. The bed beneath me feels like a dysfunctional hoverboard: my body pitching left shoulder down, right toes up, and then right shoulder up, left toes down, as the first real waves make the ship roll. Grope around for my cell phone, which I've got lodged between the mattress and the wall. *2:40 a.m.*, the screen reads. Beyond the curtain, the sun still gives a glow. I turn over slowly, trying not to wake Carolyn in the lower bunk. We have settled into an easy routine of mostly ignoring each other. I can tell she doesn't want me in her story, and that's fine; I prefer that our cabin remain a kind of sanctuary, where interaction with others isn't required.

At breakfast, the galley is empty save for Rick, who just finished his watch. Each of the three mates rotate through the bridge in four-hour shifts. Rick will steer from four to eight, in both the morning and the evening, every single day, for the next two months. I walk over to the table along the farthest wall, where he and the rest of the crew tend to congregate. As in most lunchrooms, there are unwritten rules here that dictate who eats where. But I opt to break the ship's social codes rather than eat alone, and sit down in the swivel chair directly across from his.

"You weren't the only one who woke up in the night. Three a.m. or so, that was right when we exited the Strait," Rick says. His name is embroidered on the right breast of his navy-blue work shirt, which he's tucked into his belted tan pants. "We're only going eight knots now, to try to let the storm pass in front of us."

A wave slaps the porthole as the ship pitches to the starboard side.

"Isn't this a storm?" I ask.

"This isn't much of anything yet," he says. "But don't worry, we're going to try to keep from getting clobbered."

"Right," I reply, tentatively polishing off my kiwi and yogurt and pumpkin seeds. Rocky, the bronze-cast penguin, looks over at me from his perch on the partition as if to say, *There's a reason the others are all skipping breakfast.*

I place my dish in the sink and walk down the hall to where the scientists have set up camp. Part high school computer room, part NASA mission control, the Dry Lab is the social and scientific hub of the ship. It's got the same teal plastic floors found elsewhere on the *Palmer,* plus long black countertops and a central information console where the weather, ship location, livestream of the back deck, and contour map of the seafloor are displayed. This flexible, utilitarian space can accommodate all the different kinds of experiments that might be conducted onboard. The bookshelves' extra bars keep binders from tumbling off in high seas, and the drill holes lining the tabletops enable users to create a customized network of drawn elastic to hold their electronics in place. Even the trash can is lashed to the map table with inch-wide webbing.

The Swedes on the TARSAN team have unpacked a suite of three monitors and three laptops on the counter closest to the door. Their goal—easily the most ambitious of the mission—is to be the first people to send an autonomous vehicle (AUV) under the Thwaites Ice Shelf. Anna Wåhlin, the team's principal investigator and one of Sweden's first female oceanographers, hunches in the corner, skimming the results of the previous afternoon's test run of the AUV. She is flanked by Jonas Andersson and Johan Rolandsson, her two technicians, who wear matching black beanies and work pants. In the far corner, Aleksandra Mazur, a Polish postdoc who studies the life cycles of icebergs, looks over the mission code. During yesterday's launch, a software crash forced the team to change computers at the last second. The internal clock on the replacement had yet to be switched over to Coordinated Universal Time (UTC), the standard to which all of the *Palmer*'s data-gathering machines are set, which meant they lost communication with the thirty-five-hundred-pound submersible for a couple of gut-wrenching hours. I ask Anna if she would be willing to chat with me, but she demurs, saying she needs to fix what went wrong before doing anything else. One of the last times someone sent a sub beneath an ice sheet, the multimillion-dollar machine did not return.

The Thwaites Glacier Offshore Research (THOR) team has claimed six computers along the starboard wall, as well as the island in the center of the lab. While the Swedes will use a sub to create a more accurate picture of warming and glacial loss in the present, those with THOR will be scouring the sediment at the bottom of the ocean for clues to how this system functioned over the past twenty thousand years. Like most of the groups on the *Palmer*, THOR's motley crew is a combination of career scientists and students sailing south for the first time. What unites them—beyond their shared passion for mud—is that, before this trip, they had never all been in the same room together.

Rob, who heads THOR's efforts, convenes an impromptu planning meeting. "Transit to the test site will likely take ten days," he says, "and many of us won't be feeling too well for some of that, so I'd like to get on the same page now about our top priorities." He leans over a laminated map of the inner Amundsen Sea. Here the contour lines measure not the elevation of land above sea level but bathymetry, the ocean-floor depth. The map appears marked by a bunch of random snail trails, each of which represents the path of a previous cruise. All of them used sonar to determine the distance between the ship and the seafloor, producing topographical information far more detailed than a satellite can provide. It is exhilarating to stand in front of some of the most comprehensive renderings of the Amundsen Sea on the planet. Exhilarating—and also humbling, since more than 70 percent of the plotted area contains no information at all.

"Is this somebody's box of rocks?" Cindy walks through a doorway separating the computer stations from the laboratory space and lofts a cardboard carton above her head.

"Nope," says Mark Barham, who is on board to service a set of deep-sea moorings.

All of the members of the THOR team shake their heads.

Kelly Hogan brushes her fingers over a segment of the map that contains far more data than the rest: there, two deep-blue channels run along the ocean bed all the way to the calving edge of Pine Island Glacier (PIG), which lies to the east of Thwaites. For a moment, it looks as if she has fallen under a spell. "I like to think of the shape of the seafloor as the ice sheet's death mask," she says. "As an ice sheet retreats, it scours the ocean bed. The marks it leaves behind can tell us a lot about how this particular glacier behaved in the past, which will help us make predictions for the future." She explains that because PIG holds the dubious honor of being the fastest-flowing glacier on the continent, it tends to attract funding for study. Rachel Clark, a PhD student, leans a little closer. "We've learned that PIG is deep and narrow," Kelly continues. "The glacier sits in a trough that rests well below sea level. But that trough has relatively high sides, which keep it contained. As Pine Island deteriorates, it's not likely to destabilize the ice around it."

I like Kelly immediately. I like the way she tucks her skinny jeans into her striped socks for warmth, her studded rocker belt and thick woolen sweater from Norway. I like the seriousness with which she trains the next generation of scientists, her long-standing friendship with Ali, and how she is quick to remind me that many of the world's seafloor maps are produced by the oil and gas industry—another reason we know so very little about Antarctica, where fossil fuel extraction has been forbidden for decades.

Over the next fifteen minutes, she and the rest of the THOR team take turns explaining to me the differences between Pine Island and Thwaites. PIG is long and thin, and high walls of rock help to hold it in place. "Thwaites is a whole different beast. It's really wide, with no significant topography underneath to pin it on. At least none that we can see from the very limited surveys that have been conducted," Kelly says, touching a map segment free of contour lines.

"Pine Island is sort of like the Colorado River—its flow is fast but contained? Like it's sitting at the bottom of a canyon?" I ask, trying to translate what I am learning into a metaphor that feels more familiar to me. "And Thwaites is more like the preindustrial Mississippi, wide and relatively unbounded?"

No one speaks; each person lost, I suppose, in their own corrective thoughts.

Finally, Ali says, "Alpine glaciers behave kind of like rivers, but Pine Island and Thwaites are ice streams, which are part of the West Antarctic Ice Sheet." An ice stream is an area of rapidly flowing ice within an ice sheet, a solid that behaves more like a liquid, especially relative to its surroundings.

"We know that, generally speaking, Thwaites is quite exposed. The trough it sits in doesn't have underwater mountains on the sides to keep its losses contained, like Pine Island. Instead Thwaites rests in a basin that deepens and broadens the farther inland you look. So that's unnerving, if you think about it too hard," Kelly adds.

"The net rate of ice loss from Thwaites is already more than six times what it was in the early 1990s," Rob says. Satellite imagery suggests that this year alone it will lose fifty billion tons of ice, or the equivalent of the Great Pyramid of Khufu eight thousand times over. Rob coughs and flips through a stack of these printed satellite images that illustrate ice coverage in the Amundsen, drawing his teammates' attention back to the task at hand. The one dated to less than a week ago shows large parts of the bay directly in front of Thwaites as being uncharacteristically ice free. "We're hopeful that we can get into this area," he says, consulting the grainy image. "We don't have any sediment cores or survey of the seafloor from here because this was historically covered by a big floating ice tongue."

I ask a couple more questions about how to interpret the maps that lie before us, but the longer I prod, the shorter the answers get. Kelly finally shoots me a look that says my inquisitiveness is getting in the way, so I excuse myself and shuffle over to my desk, in the Dry Lab's farthest corner, by the broken printer. For the rest of the morning, I ready my workspace for the high seas ahead. As

I screw in a bunch of eye bolts salvaged from a drawer of miscellaneous metal parts, I think about how, as I prepared for this journey, I tried to calculate the impact it would have on both the climate and the ocean-ice interface my shipmates study. One estimate suggested my round-trip ticket to Punta Arenas would pump 5.3 tons of CO_2 into the atmosphere, ultimately resulting in the loss of sixteen square meters of sea ice. Which means that the approximately 355,000 gallons of diesel the *Palmer* will burn through getting to Thwaites and back will melt almost eleven square *kilometers* of the ice we are here to study and, maybe, to protect.

I am ashamed to admit that despite understanding the negative environmental impact of the mission, I still want to be part of it. Am proud to be part of it. A similar feeling surrounds my desire to have a child. As a white woman born in the wealthiest country in the world, I know that all the things I do—love, play, work, and potentially parent—actively break down the web of life upon which we all depend. Should I have a child, their greenhouse gas emissions will cause roughly fifty square meters of sea ice to melt every year that they are alive. Just by existing, they will make the world a little less livable for everyone, themselves included. This weighs on me, but my desire to make a family persists.

When I try to understand why that is, when I try to think beyond the simple arithmetic of the carbon footprint, whether it's in traveling to Thwaites or in having a kid—when I justify these decisions in terms of the potential good they create: the stricter environmental regulations that might result from our data and our stories, for example, which could significantly lessen my child's impact in the future—I end up not just sounding defensive, but feeling defensive as well.

What reckoning, I wonder, *am I trying to keep from myself? What delusion do I wish to maintain?*

As I thread elastic medical tubing between each of the bolts, preparing for the twenty-foot-high swell ahead, I think about something Elizabeth Kolbert wrote in a 2012 *New Yorker* article,

"The Case Against Kids." She says, "When we set the size of our families, we are, each in our own small way, determining how the world of the future will look. And we're doing this not just for ourselves and our own children; we're doing it for everyone else's children, too." Kolbert, a mother of three, suggests that the decision between no children and "yet another child" ought not just to be about how many diapers you want to change or the care you hope to receive as you age, but about how increasing the number of human beings on earth decreases the planet's ability to sustain all life.

It is an argument whose long roots can be traced from the present-day interest in population control back through the movement for voluntary human extinction, to biologist Paul Ehrlich's 1968 book *The Population Bomb,* and finally to the work of the nineteenth-century economist Thomas Malthus, who argued that famine and war occur when population growth outpaces agricultural production. The fundamental premise is clear enough: the earth would be better off without us or, at the very least, with fewer of us. But too often this logic has been used only in select cases—to restrict poor people of color's right to regenerate, while the fertility of white women, like myself, isn't directly targeted—which is one big reason it doesn't sit well with me. The other, related reason is more difficult to describe.

Right now, it is frightfully easy to accept, without interrogation, that the world is ending. But there isn't just one world, and this isn't the first time much has been lost. Having children can be an act of radical faith that life will continue, despite all that assails it. In her essay "m/other ourselves," Alexis Pauline Gumbs writes, "To answer death with utopian futurity . . . [is] a strange thing to do. To name oneself 'mother' in a moment where representatives of the state conscripted 'Black' and 'mother' into vile epithets is a queer thing . . . we were never meant to survive and here we are creating a world full of love." For Gumbs, mothering is action: it both resists historic violence while also creating the context for other futures. "The radical potential of the word 'mother' comes after the 'm.' It is the space that 'other' takes in our mouths when we say it . . . all day long and everywhere when we acknowledge the creative power of transforming ourselves and the way we relate to each other." While I can't lay claim to Gumbs's revolutionary spirit, informed as it is by centuries of reproductive injustice suffered by Black women in the United States—an injustice partly fueled and directed by ideas like Malthus's and Ehrlich's—I can celebrate the idea that to have a child means having faith that the world will change, and more importantly, committing to being a part of the change yourself.

Soon I start fussing with the bungee cords the previous occupant ran through each of my workspace's six drawers to hold them shut when the ship rolls. Some knots are more difficult to untangle than others; I sit there with my fingers working the fabric-covered strands and my thoughts circling around one word, *want*. I want to have a child. Want family suppers and weekends walking in the woods together. Want to bathe, both of us in the same warm water. Yes, that. And also something even more fundamental. Want to share this body my mother made for me. Want to care wholly for another, to receive the gift of giving away everything imaginable.

Where is the computer program that can calculate the heat sequestered by humility? That will account for the fact that my child will not be steeped in exactly the same myths I was—those self-serving fantasies about how it's my right to inherit the whole earth? My child's beliefs will form in two languages and will rise up out of two different places. Their origin story will be more layered, more contradictory, and therefore, hopefully, more flexible than mine. The illusion of stability will likely fall away during their lifetime, no matter what I tell them about who they are, where they come from, and where they are going. And where is the model that accounts for what a tremendous responsibility it is to become a parent now—to know we must act, with both our children and more than our children in mind? I worry the knotted strands in my hands. Might satisfying some desires strengthen—not diminish—our accountability to one another? This possibility is, I suppose, why I signed up to travel to Thwaites in the first place.

ANNA:

I don't know how this crew will act toward a female PI [principal investigator]. That remains to be seen.

JERMAINE:

I just found out my wife is pregnant. A second baby is coming. I don't yet know if it's a boy or a girl.

MEGHAN:

I was born two weeks early, but my mom knew. She told her boss, *I'm going to have this baby today.* They were like, *You're a first-time mom, live your life, the baby's not coming yet.* She was living in North Carolina, doing water testing for some company, I think. My dad was in grad school, stressed, trying to write his thesis. My mom says she spent the day doing her taxes, like, trying to finish all this paperwork because she knew she wouldn't have time soon. She kept telling everyone, *I'm going to have this baby in a couple of hours.* But no one listened to her. She always talks about how frustrated she was, like, *I know this, I know this; I know women have been wrong in the past, but I know this.*

RACHEL:

I don't know if it's scientifically accurate, but whenever I'm seasick, I eat gummy bears and I feel better.

PETER:

The stuff that we don't know about Antarctica is staggering, completely staggering. At conferences, from time to time, they show these maps of ship tracks around the world, and the Southern Ocean is basically empty.

GUI:

I have every episode of *Monty Python*. I don't know if I'll finish them all on the cruise.

ANNA:

I study the ocean around Antarctica because I'm curious, because we still don't know how it works and that bothers me.

BECKY:

I enjoy not knowing sometimes.

JOEE:

You really don't know until you get there.

"Go out and buy a new set of pajamas," the senior communications manager for the British Antarctic Survey told me during a precruise conference call. "There's a good chance you're going to get really sick crossing the Drake Passage. And I always find it easier to ask for help from strangers if I'm looking at least halfway respectable." After I hung up the phone, I headed to the nearest shopping center and purchased gladiola-covered cotton pants with wide lavender cuffs, plus a backup pair with blue butterflies. When I cheerfully told my husband what they were for, he just looked at me with a mixture of pride and bewilderment, as though I was the only person he knew who could get excited by the prospect of revisiting their breakfast in twenty-foot seas.

Now, as the storm intensifies, fewer people appear in public spaces, and those who do are indeed wearing pajamas. In the Dry Lab, Bastien leans back in his desk chair and props his fur-lined L.L.Bean slippers on the countertop near my computer. Gui plays Tetris on his phone. Peter hops up on the counter, presses his back against the exterior wall, and declares, "At least I feel less sick in here than I do in the galley." He takes a tentative sip from his bright-orange mug, which he brought from England along with the Yorkshire Tea it contains.

I look out the porthole just above Peter's right shoulder and momentarily see three slate-gray petrels clinging to the waves, pirouetting in the troughs between peaks. Then the window pitches seaward again. A wave washes up and over it, giving me a glimpse of the bright blue bowels of the Southern Ocean. One second it is as though I were peering into a washing machine on rinse, the view lush and frothy and wet, and the next I see mostly sky.

"The creaking," Peter moans. The stressed steel hull sounds like a rope under tension. I fold up a few sheets of paper towel and place them around the leaky porthole just above my desk, to try to keep saltwater from dripping on my computer. That morning, I stood on the bridge wings and watched the ship's bow split the ten-foot swell, the air heaving and heavy as the sea, sure that whatever weather swirled before us, it would only be worsening.

Across the hall, in the Electronics Lab, a meter-long monitor tracks our progress. At the bottom of the screen, the Antarctic Peninsula reaches out like an octopus tentacle in search of prey. The *Palmer*—represented by a glowing blue triangle—hovers about half a centimeter off the tip of Tierra del Fuego. The distance between it and the great southlands remains vast. How long ago was it that we left? Three days? Five? Looking at the little two-dimensional icon, I wonder whether the ship is moving forward at all.

"We have a really long way to go," I say to Barry Bjork, the electronics technician, who is hanging an oversized crossword puzzle on the wall.

"I know," he says. He picks up a gourd and takes a sip of maté, then steps back to regard his work. "My dad got me started on the puzzles about twenty years ago. I didn't take to them right away, but on the ship they're an event. You'll walk away and come back, and the one you were stumped on will just come to you. Or someone else will have written it in."

Together we laugh at the clue for 38-down: Who once described Puritanism as "the haunting fear that someone, somewhere, may be happy"? Neither of us knows the answer. Barry tells me that he used to sell and install electronics on private yachts, but the 2008 crash sent his business belly up. He's been working part-time on government vessels in Antarctica ever since. "We're in the crossing now, so there's not too much else to do," he says. Then he adds, more quietly, "My dad, he passed just a few days ago, on the thirty-first."

"I'm so sorry," I say. We set sail on the twenty-ninth, which means the rudder trouble sent us back to port on the day his dad died. Barry pulls at his gray ponytail, then tucks his hands into the pockets of his worn green Carhartts. I tell him that I will return, then take the stairs two at a time to get to my room on the deck above. There, I root around in my candy stash, searching for something extra special: a stone-ground chocolate bar wrapped in gold tinfoil. It feels like a completely inadequate gesture, but nothing else comes to mind.

"Hopefully this can bring a little sweetness into these bitter days," I say as I hand him the chocolate, unable to imagine having to miss your own father's funeral. Barry looks at me, bewildered, and I wrap him in an awkward hug.

"I like the puzzles because they bring people together," he finally says.

GEORGE:

I came to the United States in 1992 from Latvia. I came as a tourist with my daughters and my wife at the time. We met a lot of good people in Flagstaff, Arizona. One of the friends I met, William J. Breed, was the head of the geology department [at the Museum of Northern Arizona]. He was in Antarctica in 1969 with Laurence Gould's expedition on the Beardmore Glacier, the ones who found those early reptile fossils. He told me, *George, you have to go there. You will fall in love.*

I was always really interested in science, but I don't do it myself. I have an engineering degree. Living in the Soviet Union—I mean Latvia—there was a slim-to-none chance of being involved in science, because it was difficult to make your way from your education to a work position. It's hard to explain. The decision to hire you wasn't based on your abilities or your knowledge. It was based on who you know and who can help you get something. I come from a very common family, so I didn't have the chance to work in science. But it always fascinated me.

I was raised by my mother and grandmother. My grandfather died very early, and I never had a father. When I married, I had two daughters, and now my daughter has two daughters. I've been surrounded by women all my life. My family and I, we waited thirteen years to become citizens. As soon as I became a citizen, I applied for a job in the United States Antarctic Program. I started at McMurdo, but soon I was sailing on the *Gould*—the ship named after the professor who had been my friend's advisor—and eventually I got transferred here, to the *Palmer.*

JERMAINE:

My first six years on the *Palmer*, I worked alongside my father, who retired in 2016.

JACK:

When I was a kid, I wanted to be a scientist. This is the closest I will ever get, unless I decide to go back to school for whatever reason, which I really don't see myself doing. Me and Julian, we've started our own science experiments in the kitchen. One day we cooked the dried fruit into the vegetable stew. We've been mixing and matching things that probably shouldn't be mixed and matched.

MARK:

Jesus, I don't know how many moorings I've flipped in my life. Quite a few. Over fifty, I reckon. But these are the first I will do in Antarctica. Apart from the ice, it's all similar stuff. The mechanics of getting the mooring [and the sensors attached to it] out of the water, into the boat, and then back into the water is all fairly similar. Most of the work is hardware based. I change out shackles and change out bolts. It should be kind of obvious. The ones that are fizzed away, those are what I replace. I put on new ones, change out the batteries, and that's kind of it, really.

GEORGE:

It's not that I've been working on the boat a really long time; I wouldn't say that. There are some people who have worked much longer in this program than me. I just have the opinion that when a person works in a position for a while, usually that is an indication that the work is good for them. My colleagues Barry and Sheldon, all of us are about the same age. I'm almost sixty right now, will be in two months. I think it's safe to say that the three of us that usually work on the *Palmer* as ETs [electronics technicians]—well, we each find meaning in what we do.

JACK:

I've been trying to ask everyone: What's your job? What do you contribute to this expedition? I've been learning a lot about the science. Like before I got on the ship, I was thinking that the entire ocean is cold, but now I know way at the freakin' bottom, there's warm water. It's a lot to take in: like eating an elephant, you have to take it one bite at a time.

DURING THE NIGHT, THE SHIP'S roll deepens. Unable to sleep, I lie in my bunk and watch the curtain. Every minute or so, the hem of the gold polyester swings out into the room, slow and deliberate, as if willed to the horizontal position by a magnet. It hovers there for a couple of seconds, then just as slowly starts dropping back down again. I am reminded of old horror movies, poltergeists, a blender turning on for no reason, an oven door flopping open and shut. Most everything in the cabin appears possessed. My steel-toed XtraTufs skitter across the floor, along with the chair I forgot to bungee to the desk and the notepad and pen I failed to tuck into a drawer.

In ship terms, this strange behavior means that we are "taking it on the beam," the boat running parallel to the waves, rocking side to side in some pretty big swell. Eventually, a little before five o'clock, I decide to shower. Launch my butt over the edge of the bunk, then lower my right foot to the next rung on the ladder. Tentatively I work my way toward the heaving floor. The ship tips to the starboard side, and I lean in the opposite direction. When I reach the shower, the water falls right out of the stall and onto the floor—until the boat heaves the other way and the problem is solved, if only momentarily. I grab hold of a metal handrail and enter the hot stream as the Eucerin tumbles from the shelf where my roommate left it, just missing the toilet bowl.

In addition to tampons, hand warmers, and glove liners, I brought along a Ziploc bag full of prenatal vitamins. The label boasts folic acid (to prevent defects in the brain and spinal column), ginger abstract (for help with morning sickness), and iron (to make extra blood). Since coming to Antarctica meant putting off pregnancy for a year, I figure getting an early start on vitamins is the least I can do. After toweling off, I pop another meclizine tablet and one of the enormous, earthy-green pills into my mouth. Steady myself against the sink and lift a palmful of liquid to my lips. The water, which has been sucked from the ocean and made drinkable through the ship's desalination system, tastes like algae.

In the hallway, the whiteboard reads: *Decks are secured. No one is allowed outside.* I fill my thermos with hot water in the galley, toss in some loose-leaf Earl Grey, then head to the bridge. Walking up the stairs is like trying to navigate a fun house—suddenly a distance that seemed like one foot becomes two, as the landing lifts with the ship. When I finally arrive at the topmost deck, I can't open the door at first. I wait maybe thirty seconds, then try again. What was too heavy to move suddenly becomes easy as air. Inside, Amy Winehouse's *Back to Black* plays softly on the stereo. Everything else onboard rattles and groans and squeaks: the bookshelves, the navigational guides, the Soviet-green furnaces, the leather chair against its metal trim, the pens and compass in the jar that will soon fall over. On the wall, a clinometer illustrates how much the *Palmer* heels, or tilts, in each wave. We rock to one side then back, the total roll measuring nearly fifty degrees off center.

The bridge, which is shaped like a *V* with the point facing forward, has windows on all sides. In them deep blue waves rear and fold, their tops turning white, then disintegrating in the wind.

"We're still running at eight knots or so, threading the needle between two storms," Rick says.

"Two?" I respond, staring blankly at the ship tracker. Our progress still appears minimal.

"The Drake is pretty unique in that respect. Low pressures just keep barreling through here," he says. "It's like a freeway: we're letting a large-sized low pass our bow, and then we'll speed up, try to cross the highway without getting hit."

Twice a day, the same ship-routing service used by the US Navy sends the *Palmer* a personal weather report. "They take really good care in guiding us," Rick says as the bow dips down and a giant wall of water rockets up. A second later, the wave crashes across the front deck, momentarily swamping the steel. "I think they're amazed by some of the stuff we go through." On the metal cabinets behind the navigation console, the latest printout swings from a binder clip. There, the first storm's center throbs bright pink.

Months before departure, I hatched a plan with an old friend to surf the Oregon coast. After checking the local swell report, he said, "I just want to show you something," and opened a web page that illustrated, in real time, wave height in the world's oceans. The planet pulsed in an array of colors: blue for waves up to five feet high, green for ten, and orange for thirteen-foot behemoths. All around Antarctica, the ocean flickered red and, occasionally, pink, a color I saw nowhere else. Red meant the waves were fifteen feet tall, and pink meant they were taller than a two-story house, taller than two elephants stacked on top of the other, much taller than any wave I had ever seen.

"That's where you're going," Tommy said, his voice a mix of awe and fear. Crossing the Drake is one of the most dangerous transits a ship can take, its treachery not caused by sirens or serpents, as sailors once supposed, but by the Antarctic Circumpolar Current (ACC), which swirls around the continent, from west to east, unobstructed. Where other ocean currents run into landmasses that splinter and slow the sea's force (for example, the Gulf Stream, which splits when it encounters Northern Europe), nothing stands in the way of the ACC. The closest thing to obstruction is the Drake itself: at just over eight hundred kilometers, it marks the shortest distance between any of the earth's other six continents and the southernmost. Here the band of water narrows, pulsing through the choke point between Tierra del Fuego and the northernmost tip of the Antarctic Peninsula, which leads to the heavy seas we're experiencing now.

"We're still about six hours from the worst of it," Captain Brandon chimes in, touching a hand to his bald head. A quiet man from Texas who runs a cattle ranch when not aboard the *Palmer*, he can always be found in one of three places: in the galley pouring hot sauce on his supper, in his cabin, or—most likely—sitting here, with his feet propped on the wooden railing and his eyes glued to the horizon, ready to offer the on-duty mate a little extra direction if needed.

I settle into the port-side pilot's chair and think back to Shackleton's famous crossing of these waters over a hundred years prior. His Trans-Antarctic Expedition had set out from England in a wooden sailboat in 1914 with the hopes of being the first people to cross the continent on foot. Five months later, the *Endurance* would become trapped "like an almond . . . in a chocolate bar" in the winter pack ice that covered the Weddell Sea. There the ship would sink, leaving Shackleton and his men no option but to drag their lifeboats over the shifting sea ice, until summer arrived and the pack opened, giving them a chance to row toward Elephant Island. Some men camped there for a second winter on a thin shingle beach, while "the Boss" and five others traveled *back* across the Drake in the *James Caird.*

In *South*, Shackleton writes, "We fought the seas and the winds . . . the uprearing masses of water, flung to and fro by Nature in the pride of her strength. . . . Every surge of the sea was an enemy to be watched and circumvented." Shackleton even dedicates his book to "my comrades who fell in the white warfare of the South and on the red fields of France and Flanders." When I first read this passage, I thought about how he constructed the ocean he *chose* to cross into a cruel adversary in a war of his own devising. It made me both sad and furious: sad at the way it drives a wedge between us and "nature," furious at the continuing impact of this post-Enlightenment philosophy on the earth and its inhabitants in the present day.

But sitting up there on the bridge, pitching back and forth like a Cirque du Soleil performer atop a perch pole, I have to admit that if I crossed the Drake in a wooden dingy, I would probably also liken the ordeal to combat. So much separates their experience from my own. Shackleton's twenty-two-foot-long vessel, which rode just two feet two inches above the water, was made of Baltic pine, American elm, and English oak; the *Palmer* is the length of a football pitch, seven stories high, and welded from nearly two-inch-thick steel. Shackleton and his men drank brackish water and ate pemmican, a concentrated mix of fat and protein formed into cakes or stew, while we have options: blueberry juice concentrate or apple punch, fried chicken or baked hake, yogurt and canned lychees or flapjacks with chocolate chips. His sleeping bags were fashioned from reindeer fur, and half had to be tossed overboard after soaking up so much water they started to sink the ship; meanwhile, I've got a dry goose-down sack and, on top of it, a government-issued polyester comforter the color of toothpaste.

The *Palmer* dips deeper than I expect, and the steaming tea in my thermos pours down my front, staining my favorite thermal top. I lift my feet up so Fernando, who recently began his mopping rounds, can take care of the few drops that made it to the floor. "This here is the appetizer," he says. "Every day we go from the bridge all the way down to the main deck." Then he turns, cleaning his tracks as he goes.

Luke Zeller, the youngest of the *Palmer*'s deck officers, stumbles into the room. "You been up here drinking my coffee," he says to no one in particular, heading toward the pot. The ship's tossing makes him look drunk as he weaves one way, then the other.

"Slow your roll," Rick responds, rinsing out his cup.

When Rick's four-hour watch is about to end, he always brews Luke a fresh batch. Together the two discuss the state of the sea and the storm ahead.

"You haven't experienced the Drake until footprints start showing up on the walls," Luke says. "You're not feeling sick?"

"Miraculously, no," I reply. The meclizine must be working.

I place my feet on the floor, tighten my core, and stagger over to the rearmost window. Across the back deck, white water tumbles, hurly-burly, crashing into the crane and the yellow A-frame, then spilling out the long oval holes at the base of the bulwarks. There I remain for a long while, looking at our boat as it navigates the wild blue sea. Though I hate to admit it, my comfort helps me to experience the Drake differently from the explorers of yore. I'm warm, well fed, more or less sure of my own survival, which is part of what makes it possible to look out at these waves the size of buildings and to think about other things: like literature, pregnancy, and also what's for dinner.

JACK:

That first day in the Drake, I had to run to the
bathroom seven times. I didn't feel sick. I would just
be standing and talking and then the food would
come rushing up. I remember chatting with Julian
about putting something on the line; I was in the

middle of a sentence, and all I felt was food rush up out of my stomach and I was like, *Hold on*, and I just disappeared. The whole time I'm thinking: *I cannot throw up on this floor. I don't feel like mopping. I don't feel like cleaning throw-up off the floor.*

JULIAN:

Yesterday was the worst seas I've been in, ever, in my six years. It was terrible. Things were literally coming at us. Stuff that never moves—for example, our waffle maker—that thing was going sliding across the counter. We had to get out of the way. We had the rice cooker coming at us. What else? The mixer. The speakers. The fridge. Everything was just sailing all over the place.

JACK:

This job made me do a lot of things that I probably never wanted to do: I'm terrified of heights. I had to get on three different flights just to get to Punta Arenas. From New Orleans to Dallas, we had some "mild turbulence." I'm just sitting there, and the plane started rocking. I had these two ladies sitting next to me. I told them, *I just want you to know, this is my first time flying. I don't know why I'm not having a panic attack yet, but I'm trying to keep it together.*

JULIAN:

Usually we just secure things. We use little pieces of sticky, puffy plastic, and that's enough. Stuff will stay, no matter how much we roll. But for some reason, yesterday it was really bad.

JACK:

It's hard when you've got pans flying everywhere.

JULIAN:

We even had to take the water out of the steam table. We didn't want none of you guys to get burnt. It was a challenge, you know, but in the end we rose to it. Chicken wings and pizza for lunch. Veggies and baked chicken for dinner.

JACK:

We were standing over there, and Julian was looking at the wings, and he asked me, *You know any wing sauce?* I was like, *Hot wings, hell yeah!* He said, *Okay, we're going with that.* I was running around here like a five-year-old, because before I left, I said, *I want to eat Chick-fil-A and hot wings, one last time.* When Julian agreed to make the hot wings, I was like, *This is the best day of my life right now.* It helped me feel a little closer to home.

JULIAN:

If you put shitty food out there and people are not happy, they get grumpy and it could spread. We aim for good flavors, good presentation, good timing. We try our best.

JACK:

It's been ten days so far, but it only feels like it's been five. I would have never thought that I would be in the middle of the ocean right now, on my way to Antarctica. It's interesting, being from Louisiana; it floods when it rains back home, but this is extreme. All this water, I'm not used to it. Aside from the rough seas, it's been pretty smooth. All y'all are nice. And I don't mean to brag or nothing, but the food is really good. Yesterday, I got on the scale in the gym and I was weighing 132 pounds. I was happy as hell. I didn't weigh 100 pounds until my senior year in high school in 2014. I'm gonna gain some weight being on this boat!

There is a little room called Aft Control, where I go when I need to be alone. It has a mural of an alligator riding a snowmobile past a pack of penguins, a big wooden chair for the crane operator, and, thanks to its giant windows, a great view of the sea. Deep into transit, I head there with my Kindle and a notebook. But before writing or reading, I sit down and look out. See a plume of misty air and think, *Whale spout*, then think, *Mistaken*. Think, *Swell breaking*, then, *Fin slicing the sea's surface*. Whatever it is, I don't see it again.

Opposite desires—to observe the last continent disassembling and to create life—set this year in motion. I am in the middle of the first passage, crossing the open water. It occurs to me now that despite the distance that seemingly separates these impulses, both are frighteningly beyond my control. Calving or carrying, nothing I do will guarantee either coming to pass. Alone with my view of the Southern Ocean, I am beset by fear. It feels dangerous to link these longings, and even more ill advised to commit both to print. I fear that pressing my want into the blank page of my notebook might jinx the possibility of its being fulfilled. Some part of me says, *That's silly, don't fret*; another responds, *You don't know anything yet*.

Maybe a poem will help, I think, as I open Ada Limón's most recent book. In "The Vulture and the Body," her meditation on fertility treatment and roadkill, she writes, "What if, instead of carrying / a child, I am supposed to carry grief?" Her bravery floors me first, followed by how she unflinchingly transmutes her inability to bear children into other forms of flourishing. The subsequent pages are full of the stuff of life: pistachio shells, sugar snap peas, colts with their tails chewed off. Part prayer, part curse, the collection suggests that the only way to survive great loss is to care for what remains with even more heart than before.

Eventually I lean back to watch the horizon pitching up and down, imagining myself into Limón's position. Recent studies show it is an increasingly common one. Among people in the West, sperm counts have halved over the last fifty years, thanks in large part to the hormone-disrupting chemicals found in everyday objects:

sunscreen, Tupperware, shampoo bottles, and fruit skins. These chemicals are linked to an increased risk of miscarriage and breast and ovarian cancer, and can cause babies to be small at birth with reproductive issues all their own, including undescended testes, early onset puberty, and fewer quality eggs. And it's not just humans who are being impacted. Marine snails, otters, trout, alligators, and even the endangered Florida panther have all been shown to suffer some form of reproductive disruption—smaller penises, lower hatch rates, even complete reproductive failure. "Simply put, we're living in an age of reproductive reckoning that is having reverberating effects across the planet," writes epidemiologist Shanna Swan in her book *Count Down*.

While wealthy people in Western nations can attempt to treat our reproductive problems with expensive interventions like in-vitro fertilization, artificial insemination, and even surrogacy, this ability to pay your way toward regeneration is not equally available to all. Within the United States, women of color suffer higher rates of infertility and yet are less likely to have access to assisted reproductive technologies. Members of the more-than-human world face even greater barriers to survival. The boat bobbles on while I think about the red-tailed black cockatoo in Western Australia, which breeds less frequently due to drought. Think about how temperature at the time of insemination determines the sex of sea turtles, and how warmer waters mean a higher female hatch rate—which, over the long term, could result in the extinction of the entire species. Think about the herbicides that have caused egg cells to appear in the testicles of leopard frogs, and how polychlorinated biphenyls (or PCBs) are likely driving the global decrease in Orca whale populations.

I can't tell if it's the rocking, the thought of struggling to get pregnant, or the idea that we've pumped the world full of endocrine-scrambling chemical compounds that finally sickens me. Ours is an age of loss. Like Limón, I hunger after that which no amount of hard work can secure, in a world of wonders we are dangerously good at unmaking.

THE NEXT DAY THE SEAS start to settle, and the whiteboard grows covered in graffiti. One message reads "THWAITES AWAITS!" Above it, in less confident script, someone has written "the thwaiting is the hardest part." In the corner, by a cartoon of circumpolar deep water melting the ice sheet from beneath, are these three ominous words "SOONER OR THWAITER." And my personal favorite: "Good things come to those who Thwaite."

The calmer waters bring a sudden flurry of activity, as everyone launches into last-minute preparations before the science starts in earnest. Members of the TARSAN team, of which the submarine work is but one part, gather in the conference room to discuss how best to use their allotted time in the Amundsen Sea. Anna strolls in late, carrying a slice of cantaloupe and a piece of pound cake slathered in Nutella. Gui, my whale-watching companion, appears in plaid shorts and Crocs. Peter just looks tired.

"I've been sleeping appallingly," he says. "Let's say, dozed. *Sleep* is a very strong word for present conditions." He pushes up the sleeves on his dark-blue sweater and collapses into an upholstered armchair near the head of the table.

Lars flashes a toothy smile that makes him look like the notorious sea leopard. When I tell him as much, he claps a hand on my shoulder and says, "I'll take that as a compliment. Because I like seals."

Bastien fusses with his pierced tragus—that little cartilage tab concealing the ear canal—before unrolling another meter-long map of our study area. Often, when an icebreaker is sent to conduct research at the Poles, the expedition does one particular kind of science. Some cruises are devoted entirely to drilling into the ocean floor for sediment samples; others might focus on creating a profile of the water column across time or space. But when you couple the Trump administration's decreased funding for the NSF with the desperate need to study Thwaites in as comprehensive a manner as possible, you get a Frankensteinian cruise like ours, where marine biologists work alongside physical oceanographers, sedimentologists alongside geologists. Each group has different kinds of data they need to collect and different methods for doing so. While the overall goal of the cruise is shared—gain a clearer understanding of Thwaites's past and present to better predict the future—the manners in which each team might achieve that goal are wholly different. Yet all must get their work done during the small window of opportunity the sea ice allows.

"Our rudder trouble ate up the two days of contingency time baked into the trip plan," says Bastien, who, I'm starting to sense, savors an obstacle. "That brings us down to a total of thirty-four science days."

"And of those thirty-four, TARSAN has the lion's share. Eighteen days," says Lars. While the two other science teams onboard study the glacial system's past, TARSAN's concerns are firmly rooted in the present. Their main objective: to figure out how much water is working its way under Thwaites, its temperature, and the myriad ways it eats at the ice above. Like a primary care physician taking an ailing patient's vitals, their aim is diagnostic.

The map Bastien's brought along is printed in four colors: red, white, tan, and black. The red dots mark the places where past cruises have measured the conductivity, temperature, and depth of the water column, a procedure oceanographers refer to as a CTD cast. The ocean is white, and what qualifies as land in Antarctica—ice shelves and ice sheets and actual terra firma—is tan. The black lines signify seafloor topography. As with THOR's map, there is a whole constellation of information in front of Pine Island Glacier and nothing from the area around Thwaites.

They suspect that Thwaites is, at least partially, following a familiar pattern of collapse. First, warm water gnaws away at the ice shelf that cantilevers out into the ocean, causing it to thin and eventually break. As the support from that shelf is lost, the flow of the glacier behind it accelerates. Over time, the grounding line—the place where ice and land meet—steps back, increasing the total volume of frozen water making its way into the sea. Melt, thin, break, retreat. But the particularities of how this process is playing out here and the other, lesser-understood mechanisms that might be supercharging it, are what scientists don't really understand because, of course, they don't have the data.

The first images of Thwaites are from Operation Highjump, a 1946–1947 naval mission, which generated a few aerial photos of the glacier. Charles Bentley was one of the first people in earth's history to stand atop Thwaites. In the late 1950s, he spent twenty-five months driving across West Antarctica in a tractor, detonating explosives and listening for the echo when the sound wave hit bedrock beneath. Bentley discovered that, contrary to what most glaciologists thought at the time, the ice that covered Antarctica wasn't a thin layer of frozen liquid riding on a slab of solid earth but an ice sheet, up to two miles thick, resting on land that the weight of the ice itself played a part in pushing below sea level. A few more images were taken through the 1970s, then, for the much of the 1980s, next to no new information was generated. Any real calculations of the glacier's retreat began in the early 1990s and were reliant upon satellite information. While it's clear Thwaites is receding at rates that are, by present-day Antarctic standards, very fast, there is currently no way to demonstrate how fundamentally these movements stray from its most recent baseline or exactly why they do.

"West of the Thwaites Glacier Tongue, that area is absolutely uncharted," Lars says.

Bastien draws directly on the printout with a red Sharpie, a wide crosshatch over this nameless bay. "Open and uncharted," he captions his addition. A shudder runs down my spine. With the honor of being the first people to visit this part of the planet comes great responsibility. The discrete amount of time we have been allotted means we are playing a zero-sum game: each investigation undertaken, each bit of data gathered, equals the foreclosing of other possible experiments and the reference points they too might yield.

After a week in the company of these scientists, I've learned that our predictions of just how quickly this glacier could come apart are even more speculative than I originally thought. Modelers often use past events to project what might happen in the future and may gauge that model's accuracy by seeing if it can reproduce current conditions. Take someone working on sea level rise in New York City, for example. If they model a future in which flooding that happened during Hurricane Sandy is impossible, then something is off with their math; they will need to tweak their model until the eastern shore of Staten Island, the Rockaways, Red Hook, and the Battery are all underwater.

But in Antarctica, not only is the observational data severely limited, the very processes that define large-scale glacial retreat remain relatively unknown. Many of the models we use to project Antarctica's future struggle to reproduce the rapid and dynamic changes we know are happening, which suggests that some other physical process—in addition to glacial melting—might be taking place. In 2015, three of the foremost scientists studying ice sheet dynamics published a controversial paper outlining the parameters for something that explains that discrepancy. They called their newly proposed physical process Marine Ice Cliff Instability (MICI). Richard Alley, Robert DeConto, and David Pollard argue that once the cliff face of a calving glacier reaches about one hundred meters, it could become inherently unstable. And if you break off a chunk of it, you might expose an even higher, more unstable cliff in the process. Like dominoes or a house of cards, when one piece goes, so does the rest. Or at least that is the idea. But because no human being has ever witnessed MICI in real time, most models, including those used by the IPCC, do not include it when predicting future sea level rise rates. "We could be underestimating the worst-case scenario," Alley concluded in a subsequent article in *Scientific American*. "But we really do not know."

Anna takes a slow and deliberate bite of melon while surveying the study area. Her main goal is to send an AUV beneath the ice shelf. The sub, should it go under, will hold close to the bottom, rendering the seafloor in unprecedented detail. "We know there are these tiny corrugation ridges littered near the grounding line of many big glaciers, and I suspect they contain clues as to how this ice sheet fell apart in the past," Ali told me earlier that day, "giving us a better understanding of whether

MICI is possible in the near future. But no one has ever been able to see them up close." At least not at Thwaites.

"Just north of PIG, there's another area we know nothing about," says Lars. Bastien marks it with his red pen. Once we enter the study area, the scientists will start running shifts like the technicians: twelve hours on, twelve hours off.

"We'll steam along the eastern margin of the continent," Anna says. "Our first destination is the Edwards and Lindsey Islands."

"Correct, correct," says Lars. "We plan on splitting science days there with Scott, who will be on land hunting for penguin bones." I feel as though I am standing in a war room during a crisis. "We have four days set aside for the islands, but we don't have to use it all in one jag," Lars continues. "We'll pass this way on our way out as well. While we're off tagging seals, the ship could be taking water column samples."

"Or we can do another test deployment of Ran," says Anna, using her nickname for the sub—a reference to the vengeful Norse goddess of the sea.

The two scientists look at each other, trying to decide whether they are at odds, or if there might be a way for them both to get what they want.

"Ran or the gliders can be deployed and recovered while the other teams are on the islands," says Bastien. "We can split the time fifty-fifty between the two. Meaning we only have to use, let's say, two of the days allotted to carry out four days of science."

"If we start splitting everything fifty-fifty," says Johan, the sub tech, "we're going to be here until the end of May."

"I think the question is how much can we do before the human body collapses," Bastien says to the group. He has been standing the entire time, bouncing from one foot to the other.

"At this point, given the difficulties we had with the operating software and getting Ran back onboard in the Strait, it's really important that we test her again," Anna says, quietly drawing a line in the sand. It seems strange that basic questions—like who is going to do what and when—are still being settled, but then again none of us knew this much of the Amundsen would be ice free until very recently.

"After all these test deployments, we'll be able to send Ran seven kilometers under the ice," jokes Johan. If one kilometer is high risk, seven equals submarine suicide.

"I need to start bringing something to hit people with," says Anna.

"Why don't we aim to create a profile of the currents moving between Burke Island, Cosgrove Ice Shelf, and the Lindsey and Edwards Islands?" Lars proposes, drawing a blue triangle on the map between the three points.

"No one has done that before," says Bastien. "We can box in the current, which would help us get a better sense of how much warm water is reaching PIG and Thwaites."

"That way we can make the most of the second test deployment, sending Ran across the opening at Cosgrove," Anna says, satisfied.

"*Nature* paper!" chimes in Peter, suggesting the endeavor might yield an article for the field's leading academic journal.

Lars laughs.

It takes over an hour of deliberation before the team arrives at a plan, broken into thirty-minute segments, outlining which machines will be deployed where and in what order. The plan spans our first two full days in the Amundsen Sea, and, though we don't know it yet, it will begin to disintegrate even before the first glider is launched.

CHRIS:

On the ship, I pay attention to so many different clocks. For instance, I think of this day as day 34, not February third. We catalog the science data according to the Julian year: January first is day 1, December thirty-first is day 365. I've got a regular calendar too. I keep track of the time in Denver, which is where my home office—the Antarctica Support Contract—is located; and I keep track of the time in Punta Arenas, which is the same as Santiago time, but not always, because of daylight savings. And then there is the "clock" that we all think about, which is: *When is dinner?* Right? Life tends to circulate around shared activities: meal times and meeting times. The clock on the wall by your desk over there—that clock is always wrong, because it's on UTC time. Laypeople often call it Greenwich Mean Time, and like the Julian year calendar, we use it to log exactly when the data is collected.

ALEKSANDRA:

The computer we use to run Ran crashed during the test mission, so they moved the mission to a private laptop, which was running on a different time, not UTC time. That's why we couldn't track her. When she's in the water, you can usually see how the mission is progressing. You can see where she is and what she's doing. But because the time didn't agree, we couldn't see anything that was happening. As soon as Ran dove, she was by herself—like, really by herself.

MARK:

On transit, when there's not a great deal going on, every day morphs into the next. It really doesn't matter. Well, it's not that it really doesn't matter; it's just that I don't know what day it is.

PETER:

You lose track. I mean, we left on—we were on the boat on—Monday, refueled on Tuesday. We were back in PA on Wednesday for the rudder problem, which lasted until late in the day Thursday. Then we left again. Friday, Saturday, Sunday, Monday. I make it seven days since we set out, but then I'm also not positive. I've never had to confront seven weeks as an unbroken period of time. Back home, you do your five days of work, then weekend, then five days of work. It breaks things up and takes on a rhythm. Whereas on the boat, that doesn't happen. You don't ever stop being on the boat, and you don't ever stop working, really. We lose those markers of time where you can go, *Oh yeah, it's Saturday; it's obviously Saturday.*

ROB:

We are three to four hours out of sync with astronomical time.

ALI:

My daughter, she doesn't really comprehend time right now. We spoke to her about my going to Antarctica; I mean, she knew I would leave. And now we have been here how long? A little over a week, and she's doing just fine, which is comforting for me to hear. She will ask where I am, and then she will get completely distracted by whatever thing she looks at next. When I tell her I have six more weeks on the ship, I don't think she has any idea of what that means.

MEGHAN:

Living through a period of such accelerated geological transition is . . . a lot. It's a lot. Figuring out how to stress that we're seeing changes on a ten-year time scale that used to take place over thousands, tens of thousands of years is the hardest part. The rate of change—that's what's terrifying, the little time it's taking for these massive transformations to unfold.

ANNA:

It's extremely frustrating to work in an area where we have so little data over time. Especially when you read in the newspapers that this part of the Southern Ocean is warming. We all raise our hands and say, *We really don't* know *that.* We don't have the observational data to prove it.

We need to start the long time series now, in order to get the data that in fifteen years can show warming over time.

JERMAINE:
I've crossed the Drake so many times I can't keep track.

FERNANDO:
If you don't work hard, then the time is slow and you're thinking a lot. If you keep busy, the time passes fast. Most of the days, I am busy cleaning, painting, or doing carpentry. Maybe tomorrow or the next day we will see an iceberg. When that happens, time will speed up, because we will all be excited and looking at it.

ACT TWO

Part One | *Into the Ice*

SETTING: Here, where certain colors begin: apricot, lemon curd, rose hip, nectarine. All are born in the jam-smudged southern sky.

Most everyone—save the engineer, the cook, and the mate on watch—sleeps, sunk deep into their bunks now that the ship finally stopped rocking. I lift just the corner of the curtain so as to not wake Carolyn. The light outside cuts like a scalpel, sharpening the edges of the clouds gathered close to the horizon. Beyond them, the sky is just this side of periwinkle—perplexing and quiet, not quite purple. How many days has it been since most of the color leached out of the world? How many days since we steamed past Desolation Island into open ocean? How many days since I spoke with anyone not on this boat? Swaddled in these strange slippages of time and space, I leave my room and head to the topmost deck.

"The *Palmer* popped through an invisible barrier during the night," Rick tells me. They call it the Antarctic Convergence, or Polar Front, the place where the cold water that swirls around the continent sinks beneath the less dense, warmer water from the north. Like a piston in a pump, this simple exchange drives ocean circulation the world over. It is, he explains, one of the many ways in which *here* and *there* are, despite great geographical distance, connected. Then Rick adds, "I've got something for you." I expect him to hand over a book of nautical terms or an atlas of the Southern Ocean, since he knows that I enjoy expanding my Antarctic vocabulary. Instead he points to the horizon. There it is, at 66° south: my very first iceberg.

Outside the temperature is noticeably colder, the sea fog cleared, the wind all but gone. I grab hold of the railing and take a few tentative steps on the catwalk that surrounds the bridge. Sixty feet below, the dusky ocean undulates like a big sheet of silk. My stomach drops. A few more steps, and I reach a small, triangular steel deck and sit down.

The lonesome berg rides low in the water. Like whipped meringue piped into a lopsided point, the whole thing lists to the right. Its closest side guttered and blue, the top dove gray. My eyes hold on to the ice, though I don't know what to do with it exactly, this scraggly, unorthodox thing. A few big rollers come through and throw themselves against the berg, spray lofting into the air. It is hard to say how high the mist reaches—forty or fifty feet?—because there is nothing else around to serve as a reference point. No other, more familiar objects against which its size might be inferred. The higher the sun rises, the bluer the sea and sky, the whiter the ice, until some of its facets burn so bright they appear blank.

The feeling falls away from my fingertips, and the distance between us and the berg closes. Then we pass it, or it passes us; I don't know anymore who or what is traveling toward whom, or why.

What I do know is that all icebergs are glacier born, the offspring of a parent stream. According to the *New Oxford American Dictionary*, "to calve" means to give birth to a baby cow or to split and shed a smaller mass of ice. These definitions—of animals and of glaciers both—describe the moment one thing becomes two. From cleaving, a flourishing, some new start. This linguistic echo has long delighted me, because it helped me think of Antarctica not as an inhospitable island at the bottom of the earth but as a mother, a being powerful enough to bring new life into the world. However, as I stare at the lopsided straggler, this slab of ice so diminished it's nearly gone, my enchantment with the idea that Antarctica's great glaciers are responding to us, to our actions thousands of miles away, by birthing bergs whose very bodies bear grave warnings seems all wrong. Because, I wonder, how bad must things get for a parent to make such a sacrifice?

Eventually I turn, take out my phone, extend my arm, and snap a selfie. In it, my smile looks forced, the iceberg almost indistinguishable from the clouds that scud along the horizon.

"Why didn't you tell anyone?" Rick asks when I bump into him nearly two hours later, on my way to the bathroom. My cheeks flush and I mumble something about how I got distracted, but I know it's only a half-truth; some part of me did not want to share the experience, wanted instead to sit alone with that first sentry from Antarctica. Instantly I feel guilty and a little hypocritical. To withhold the iceberg from my shipmates makes me into its sole mediator, an instinct—though I don't want to admit it—that's not very far from the impulse to name it after myself.

So instead of going to the bathroom, I turn around and tell everyone working in the Dry Lab that we have finally entered Antarctica's icy reach. In minutes, the back deck is cluttered. The first few bergs we pass are small, like the one I saw earlier. Then, on the horizon: a monolith sheared from the face of a faraway mountain range. Captain Brandon decides to give us a good show, changing course to pull the *Palmer* extra close.

The nearer we draw, the stranger the shape becomes. Its surface, vexing. The upper portions are the texture of crumpled wax paper, all sharp corners and sleek surfaces. Where the sea smashes against the base, however, the ice seems as smooth as buttercream frosting. As we draw around the backside, a milky-aquamarine spire, like a hoodoo at the top of an eroded canyon wall, rises precipitously. It's striated at the top, over half a century of snowfall and summertime warming recorded there in alternating bands. Another bit of ice, whose long, low rampart has the pearly luster of kyanite, rests mesa-like behind the first. In the ocean a ghostly form many times larger than the ship connects the two.

"This is one of the most beautiful things I've ever encountered," Gui says.

"There are loads of them," Peter adds, delighted, looking farther south. "*You* get an iceberg, and *you* get an iceberg," he says. He mimes plucking each one off the horizon and handing them out as presents.

Jack appears in his chef's uniform. "Julian told me I had to come check this out," he says.

For the next couple of hours, most don't work. We walk from one side of the deck to the other, from the stern to the bow, leaning over railings, flinging our attention outward—in a ballet whose movements are dictated by ice, by how raptly it holds us.

FILIP:

I was sitting in my cabin, doing some work on my computer. I haven't been able to do much because we had a lot of submarine prep in Punta, and then with the rocking and all, I got pretty sick. Finally, yesterday, I started working. Then I look out the window, and there's this huge iceberg out of nowhere. I had no idea we were in iceberg waters. It was thick on one side and thinner on the other; it looked kind of like you could ski down it and do a jump off the edge.

MEGHAN:

Around midnight, the sun almost went beneath the horizon. The water was very calm, so it was super reflective but also super dark and flat, which was nice to see on its own, after the Drake. I'd stayed up watching *The Princess Bride*. My phone was dying, and so I was going down to my room to charge it, but I figured: Once you're in the stairwell, you might as well go up to the bridge, right? And when I got up there, it reminded me of Utah. The icebergs were like buttes and plateaus, all the way to the horizon.

JACK:

It's nice to go up to the bridge and to see it snowing, like legit, big snowflakes. And then there are the icebergs. They remind me of that movie *Happy Feet*. I mean, I would have never thought when I was in elementary school that I would be cooking for fifty-seven people on a boat surrounded by icebergs. Would have never thought.

LUKE:

I remember seeing my first iceberg. I really didn't know what to expect, so I broke out my camera and took a bunch of pictures. And then, of course, I went right back to ice navigation. Back to focus mode. You have to always bring your attention back to the radar and to what is happening right in front of the ship. Even though you're gonna see a ton of bergs, still every single one is unique. They all have their own characteristics. Take that one over there, for instance—it looks like a swan. I've probably seen thousands, but it never gets old. And it's cool to see new people's reactions. It gets me excited to see them so excited.

FILIP:

It feels like we are actually getting there. To Antarctica. When you're flying, you just get on the plane, and then a few hours later you're halfway across the planet and you have no conception of how far away you are from where you started. But this feels totally different, like an expedition. I like that it's taken us a while to arrive.

Less than a year after I return from Thwaites, I meet Diana Richards, a freshman at Vassar, whose end-of-the-semester project on econatalism connects reproductive decisions made today to long-term ecological health. "Many young people see having a child as bad for the environment or do not wish to raise a child in a world beset by climate catastrophe," she tells me in a measured voice. The poster hanging on the wall behind her contains two infographics. One registers the different reasons young adults are having fewer children. I learn that anxiety about the environment is more impactful than a partner who doesn't want kids but less impactful than anxiety about the economy—a bleak portrait of how today's college students feel about the future they are inheriting. The other chart evaluates the efficacy of individual decisions on overall carbon consumption.

"Having one less child is more effective than giving up your car, flying less, recycling, and upgrading all your incandescents to LEDs," Diana says, pointing to the different-colored bars rising like skyscrapers from the base of her graph. "Want to lessen your carbon footprint? Go kid free."

I pop a cheese cube from the buffet into my mouth and chew, nodding. None of this is news to me, and yet I find myself destabilized hearing it confidently presented by this young woman. Still, I tell Diana that I've long been skeptical of the cold logic of carbon calculus, sure that there is a categorical error at its heart. Child-rearing is not just another consumer choice. Having a kid is not the same as flying to France on vacation or eating sirloin for dinner; it's not the same as purchasing a Prius or a Chevy Bolt. Diana looks at me as though I am not who she thought I was.

"My mom tells me not to take any irreversible actions today," she says, tucking her black turtleneck into the band of her mustard-yellow pants. "That I may want kids in the future and just not know it."

I shrug, trying to ignore the sting of suddenly being aligned with an older, more conservative demographic.

"Giving up on reproduction isn't the solution," I finally muster. The words, as they exit my mouth, strip me bare. It is the simplest way I have of saying what I have struggled to own for so long.

"Not *the* solution," Diana corrects me, "but it could help?" Looking at her, I think of a younger version of myself, the person who converted all of her bulbs to LEDs, who recycled and composted, carpooled and biked. Who watched her friends install solar panels and purchase electric cars, travel less and upcycle. But none of it stopped the earth from warming. *What if I also give up what I have been told is one of the most meaningful human experiences? Might that sacrifice be big enough to make things better?* This is the question I imagine buried in the bedrock of Diana's searching voice.

I desperately want to hug her then, because I know what it feels like to fear that there might not be many meaningful strategies left. Instead I tell her how much I enjoyed our talk, then turn to the next poster. But the ease with which she tied choosing not to have children directly to environmental care haunts me all afternoon.

It is an uncanny reversal of the kind of dictum that used to shape the discourse, the old do-it-for-your-children line. Between the birth of the environmental movement and today, the conversation has evolved from a paternalistic, stewardship-based model of ecological activism to include questioning the very ethics of bringing more people into the world—asking not "How do we save the planet?" but "Should we even be here?" Of course, back in the dawn of Earth Day celebrations, few were talking about parts per million and carbon footprints; the ice caps might have been melting, but that had yet to make the nightly news. Both our impact on the planet and our awareness of it have grown exponentially since. But the distance between these two rhetorical positions, each outlining how to care for the more-than-human world, is so great that it makes me wonder what else, in addition to the climate, has changed.

NOON THE NEXT DAY, BUT it could just as easily be midnight or six in the morning because the light has bled almost evenly in every direction. The sea: many-toned white all the way to the pale horizon. The boat plows deep into the drift—that broken turtle shell of shifting plates, a chaotic mix of flat floes, mid-sized bergy bits, and smaller chunks of brash—seeking out open water where possible and otherwise pinballing between large slabs of ice. Sometimes there is nothing that resembles a lead, so the ship must make one. The cracks never appear dead center but instead fracture outward—radial, dendritic—from the place where the *Palmer* pushes forward, its path an interplay between the ice's movements and the mate's.

I'm standing on the bow and almost warm, except for my hands, so I try a trick Kelly taught me: hang my arms at my side, extend my fingers in a fan, and pump my shoulders up and down. Blood shunts into my digits, causing a prickling feeling, followed by the return of sensation. Off the starboard side, a lone Weddell seal sashays across the white field. Its head and tail seesaw, propelling its fur-covered form forward, until it slips off the ice's low edge into the sea. Then *Palmer* smashes into the spot where the seal lounged just a moment ago. The hull groans and screeches; the boat slows, its engines chewing through even more diesel, just to inch us up on top of the floe. Soon dozens of chunks of ice rocket skyward. The water around this rubble roiling and burbling, electric blue.

For hours I stand in my cold-weather gear, watching the pack open and close around the ship. I am transfixed by the ice's labored music: the clicking and popping as pressure ridges heap up, the sibilant cascade of water draining from an overturned floe, how beneath it all I swear it sounds as though a whale were singing. Yet I have no idea how any of this—the thick rind of sea ice, the single-celled organisms lounging on its underside, the krill and seals who sup there—came into being. When I sit beneath the red spruce in my backyard, I know that it began with a pinecone, suffuse with seeds that the chickadees eat. I know something about the seasonal wobble that coaxes the seed from its protective case when the weather is right, for I have lived with spruce trees all of my life. But here I have no referent. No way of teasing apart, intuitively, the processes that gave birth to this surreal Seussian landscape.

Google is my go-to when searching for new information, but the internet doesn't reliably reach the bottom of the planet. This leaves me with two alternatives: the people on the boat or the books it contains. I'm feeling reclusive and willful, so I tromp up to the bridge and over to the bookshelf where the *Palmer*'s practical library lives. What we know about Antarctic navigation is all here: the seventh edition of *Antarctic Pilot*, *Ocean Passages for the World*, *Handling Ships in Ice*, and the National Geospatial-Intelligence Agency's *Sailing Directions (Planning Guide & Enroute) Antarctica*. Some are bound in leather, others in three-ring binders. I start with the largest of the lot, the *American Practical Navigator*. Scan the table of contents and flip to chapter thirty-four: "Ice in the Sea." It begins with what amounts to a grave warning, at least as far as the cool language of guidebooks goes:

> Ice is of direct concern to the navigator because it restricts and sometimes controls his movements; it affects his dead reckoning by forcing frequent and sometimes inaccurately

> determined changes of course and speed; it affects his piloting by altering the appearance or obliterating the features of landmarks; it hinders the establishment and maintenance of aids to navigation . . . it produces changes in surface features and in radar returns from these features; it affects celestial navigation by . . . obscuring the horizon and celestial bodies either directly or by the weather it influences, and it affects charts by introducing several plotting problems.

A handwritten maxim taped to the shelf above the map table communicates a similar message, albeit in far more economical language. It reads "NEVER FORGET: THE ICE IS TELLING YOU WHAT TO DO AND NOT YOU ARE TELLING THE ICE WHAT TO DO."

"How close we get to Thwaites depends on what the ice allows," Luke says after I ask him how he interprets the sign. Then he gets real quiet and moves a lever that makes the thrusters run in reverse. The boat backs up. Ten meters. Twenty meters. Thirty. He switches the lever and the engines kick up an octave. We lurch forward, plowing into the floe in front of us a second time, hoping to advance the little crack our last collision created. "Tricky," he says, shaking his head.

Our forward progress all but stops. On the horizon: spires and pinnacles. This morning, the floes were flatter. Now they have tented up into little peaks and mountain ranges.

"Are you nervous?" I ask the captain.

"Do you get nervous doing your job?" he replies.

A two-toned satellite image of our study area hangs behind him. The Sentinel composites come in every couple of days, giving us a vague approximation of the state of the sea ice. In them, the ocean is black. The ice in various shades of gray: darker areas denote thinner ice, while anything more than a couple of feet thick appears nearly white. It is impossible to tell, based solely on this remote sensory data, the difference between a floe that is three feet thick and one that is fifteen. For the *Palmer*, however, that difference is crucial. It is the difference between ice we can break and that which will force us to turn around.

"The satellite images give you a sense of where you want to go," Luke says. "But they don't help you get there."

"If we can punch through this section, the going should get easier again about one hundred kilometers farther south," the captain adds, referring to the patch of seemingly open water the most recent Sentinel registered. On the monitor to his right, the long arm of the Furuno radar spins clockwise, revealing any object in a twelve-kilometer radius. It looks like we are riding through the heart of a constellation dense with stars, the ice glowing in that atomic green of early computer games. The Knudsen CHIRP 3260 sub-bottom profiler has also been turned on. The technology is similar to what bats use to navigate in the dark: the transducer sends out a pulse and then listens for the echo of the sound wave when it bounces off the bottom. Halve the time between the ping and its response, and you know the ocean depth. Log this information continuously, and a rough image of the seafloor appears: full of valleys and canyons and outcroppings.

"See that little bit of open water to the right of this big floe?" Luke says, gesturing. "At least that means the ice has somewhere to go." We are inching forward, hunting and pecking through the ever-changing drift. Our course dependent upon connecting what we can see from the bridge to whatever the Knudsen and radar screens suggest.

Luke backs the boat up a third time.

"C'mon," he whispers. "Crack."

The *Palmer* grinds forward, riding up high onto the floe. Suddenly the ice yields—a little lead opens—and the sea pours in, a bright antifreeze blue.

Satisfied, I turn back to the book most sailors refer to simply as "Bowditch." Originally penned in 1802, Nathaniel Bowditch's *American Practical Navigator* is the mariner's equivalent to the Gideon's Bible. It's carried onboard every single US-commissioned vessel, but unlike the book in the nightstand at the Marriott, this tome is regularly referenced. Since its first printing over two hundred years ago, the *American Practical Navigator* has been reissued more than seventy-five times, each edition updating what we know about how to travel the ocean. It contains some of everything, from the practical mathematics necessary to calculate lunar distances to the importance of Ekman transport, a mechanism that moves the top layer of the water column at a ninety-degree angle to the wind.

The earliest editions contained the word *ice* only once, in the glossary, under the entry for *bow grace*, a chain or cable hung over the front of a wooden ship in very cold weather. Now there is a whole chapter devoted to the stuff. As I read I learn that as water molecules lose energy, they cool and contract, transforming from liquid to solid. First they form frazil ice, a flotilla of small, needlelike crystals suspended in the top few centimeters of seawater. Next comes grease ice, the soupy layer that emerges as frazil spicules coagulate, and then shuga ice, spongy white lumps that bob and sway in the soon-to-be cold-hardened sea. Nilas ice, an elegant elastic matte, fans out after, and from its rind, pancake ice is born. *This explanation*, I think, *sounds like a recipe from* The Joy of Cooking.

We are surrounded by floe, which follows after the pancakes exude salt and fuse together. The uppermost layers of floe ice never quite settle, are instead weathered by wind-driven snow, forming land features similar to those found in a desert: slip faces, long humped dunes, and arêtes. The endless expanse we plow is seemingly a single color—white—but within it a thousand variations in the key of blue. Baby blue of seal tracks; stone blue of sky's reflection. And in the space between the floes, where the nearly frozen water shows, a turquoise so deep it torques all it touches into something new.

I love the sound of the word, how its homophone, *flow*, suggests the opposite of stasis. Down below, most are stepping into the flow of Antarctic science. In the Baltic Room, Lars prepares for fieldwork by zip-tying sensors onto a machine that will be dropped over the side of the boat. Submerging them will test whether each is working and properly calibrated, something you want to be sure of before epoxying them to the foreheads of sedated elephant seals. Jack cooks up some Louisiana red beans and rice and places it on the buffet, alongside the lettuce that has lost its crunch. Lindsey checks the survival kits that will accompany the teams deploying to the islands the next morning. Every waterproof sack weighs about forty pounds and contains space blankets, emergency rations, sealed satchels of water, tents, sleeping bags, and mystery novels to stave off boredom. "Because," as Lindsey says, "when a storm kicks up down here, you never know how long it will last."

LINDSEY:

We're not going to teach you how to survive on the
islands. We're going to teach you how to open that
bag. Use this stuff if necessary. It doesn't have to be the
ultimate extreme emergency; just use what you need.
We're going to send a recon boat to find a landing site at
6 a.m. But you, too, should be ready to go then, because if
it's clear, we'll load you up and launch.

CARMEN:

Come wearing waterproof foul-weather gear. Don't be the
person skiing in jeans.

LARS:

Today we made a box with all the gear we need to catch
and sedate the seals. The syringes, the darts. Blowpipe
tube. This is the step-one box. Step two is when the seal
is sleeping and we put the tag on. We're now filling that
box with gear: glue cartridges, tags, transmitters, and
sensors. We have boxes in boxes. Bags in bags. It's like
color-by-number how to catch a seal.

SCOTT:

We're going to the islands to gather samples for cosmogenic and carbon dating, to try to get a sense of how fast this land rebounded during the most recent deglaciation. The big Ziplocs are for big bones. The smaller ones are for shells, fur, bone fragments. That makes it sound kind of creepy, right? The canvas bags are for rocks.

MEGHAN:

We're going to fill this thermos with boiling-hot water and, if we need to, pour it on the ground, because sometimes the bones and rocks are so frozen in place that it's really hard to dig them up and pick them out.

SCOTT:

In most parts of Antarctica, were you to pee or poo, it wouldn't go anywhere for, like, hundreds, maybe even thousands of years. So pack in, pack out is brought to an extreme. If you've got to pee, it's into a bottle, and you dump that bottle into a bigger bin at the end of the day. If you've got to do your number two, it's usually in a five-gallon bucket. Here, you can also go in the tidal zone. Dilution is the solution.

LARS:

Remember Lindsey's talk about small-boat safety and using the Risk Card to analyze probability and fatality? A seal bite is in the red area. It's really dangerous, and it's also quite possible, so it's something you have to think about. It's not something you deal with when it happens. You have to think about it beforehand. Don't go to a seal until it's sedated. Simple as that. If we can, we just put a blow dart in at a distance and wait: that's the best scenario.

GUI:

We talk a lot before we go to a seal. Like, a lot. The amount of times we talk about the same things—go over the steps in the procedure, each one—is, like, well . . . let's just say we talk about it a lot.

LARS:

After every seal, we also talk about what we did. What went well and what didn't. You have to come down. You're full of adrenaline. You have to come down and then move to the next one. And start the process again with box one.

ROB:

Regarding the next twenty-four hours of science that we've got planned: probably all of our best estimates will prove to be incorrect.

LARS:

Very often when you tag seals, it's a small group in a small place, so you have to get comfortable with other people's body functions. It takes respect and trust, weirdly enough. Sometimes you say, *What happens in fieldwork stays in the field*. Like, sometimes you have to take a crap on the beach, and that's something you wouldn't do in front of your colleague when you're back at work. Sometimes there's the postdoc, having a dump off the side of the boat.

GUI:

Why do I feel like Lars is preparing me for something?

LARS:

Oh, come on. We'll have a really good breakfast, then we'll have a *big* coffee, and then we'll go tag some seals.

SHORTLY AFTER SPEAKING WITH DIANA Richards, I read Meehan Crist's essay, "Is It OK to Have a Child?" and encounter two sentences that fundamentally rearrange my thinking; two sentences that, in the years since the article's 2020 publication in the *London Review of Books*, reverberate through many others' conversations about childrearing, climate change, and what ought to be done. She writes, "It was BP [British Petroleum] that popularised the idea of the personal carbon footprint. They introduced it in a 2005 US media campaign that cost more than $100 million per year, deflecting responsibility for combating climate change onto the individual consumer." Immediately I open a new tab and do a quick Google image search. It leads me to the web page of an independent art director who worked on the now-famous "Beyond Petroleum" campaign in the early 2000s. The design is deceptively simple. Each advertisement looks like the notes a college student might take: a short statement or question in black type, partly highlighted yellow, and a paragraph beneath unpacking its meaning. "What on earth is a carbon footprint?" a full-page newspaper pullout asks. The answer: "Every person in the world has one. It's the amount of carbon dioxide emitted due to our daily activities—from washing a load of laundry to driving a car load of kids to school." Then the ad invites readers to BP's website to calculate their lives' negative impact on the planet. At the bottom, BP's new logo blooms green and yellow like a many-lobed chrysanthemum.

BP took out two-page ads in the *New York Times*, the *Wall Street Journal*, the *Washington Post*, and many other major newspapers. They pasted fliers at intersections in our largest cities. Ran thirty-second spots on the big networks, and made ads that aired before selected video content online. Within a couple months of launching, BP's carbon calculator had over a million hits. In other words, the narrative that individuals are responsible for both the climate crisis and slowing its acceleration via different consumer choices was crafted and drilled into us by one of the highest-emitting companies in the world. My astonishment that BP (which has pumped more than thirty-four billion tons of carbon into the atmosphere since 1965, and which has made over $332 billion in profit since 1990 alone) designed and disseminated the tool that so many of us, myself included, refer to as a matter of course immediately morphs into rage.

Rage that any young person might find their plans to have or not have kids shaped by corporate maneuvering. Rage at the way carbon calculators suggest all life should be viewed through the warped lens of an extractive economic system where taking is assumed, with no giving, tending, or mending in return; rage that this line of reasoning fails to imagine humans as capable of making radical change, of creating a world where we are no longer defined by what we buy. Rage at how it isolates each of us, refusing to recognize what can be achieved when we work with others toward a shared goal. Rage at what it denies parents who need the transformation of our energy systems now, so their children might face fewer risks in a climate-changed world. Rage that my own desire to have children has always, for as long as I can remember, filled me simultaneously with joy and fear and guilt. Rage at the time I lost feeling ashamed for wanting to become a mother.

I can't close the website right away.

Instead I stare at it, impressed by how ubiquitous carbon math has become in the two decades since the campaign launched. I open more tabs to confirm my hunch. Discover that many environmental organizations—like the Nature Conservancy, Environmental Protection Agency, and Conservation International—now host carbon footprint calculators on their websites. UC Berkeley, Stanford, and the University of Washington all coded their own versions with distinct multiple-choice questionnaires and infographics. The Knight Foundation even supported the creation of one tailored toward grade school kids. At the end, it tells children how they stack up against others and encourages specific behavioral changes, like turning off the lights or taking fewer showers. Many of those geared toward adults allow users to purchase their way out of guilt. Just type your credit card information into the glowing interface, click *send*, and forests will be restored, solar panels purchased.

As I click through the many windows I have open—the false emancipatory promises they each contain—I am reminded of something Blythe Pepino, founder of the BirthStrike movement, said a few weeks prior, when we spoke about her choice to go child-free. When she began to participate in direct-action campaigns, she knew she had to put off pregnancy. What started as a practical concern eventually became a form of political dissent. By refusing to bear children—when population growth has been central to the success of every single modern nation-state—BirthStrikers hope to pressure elected officials to finally act with the interest of all future beings in mind.

"You know what really gets me?" she asked. "In all the conversations we have around these topics [reproduction and climate change], activism is never encouraged. But if a hundred of us manage to close down a coal plant—just your average coal plant—for one day, we also reduce the output of CO_2 into the atmosphere by sixty tons." This, she notes, is what the most widely referenced study suggests will remain sequestered if someone in the developed world has one less kid. "BirthStrike is about creating a different narrative. It's about building solidarity. People think we're opposed to motherhood, but that's wrong. We see ourselves as would-be parents, as closely aligned with mothers and caretakers. Motherhood is with us all the time." To be with motherhood does not mean privileging heterosexual relationships or traditional family making; it means working on behalf of the right of all beings to regenerate. It means making central to one's position in the world, the possibility of that world's continuation.

On the ship, I pull my waterproof bib and windbreaker over my pajamas. No tea, no coffee; I just stumble down to the mudroom and grab a fluorescent-orange float coat, zip it up, and clip the waist buckle. Lift a hard hat from the hooks. Spin the plastic dial to cinch it tight over my wool beanie. Walk down to the Wet Lab, where I pause in the doorway to the back deck to listen: no hydraulics on yet, no danger of cables snapping.

I had debated whether to wake for this sub launch, scheduled for three thirty in the morning. The first, in the Strait of Magellan, had been relatively hands-off, and the thought of sleeping through the night was tempting. But, I'd reasoned, when I return to dry land, I might soon have to endure many sleepless nights and very early mornings. *Maybe Antarctica will teach me how to become a mother,* I'd scribbled in my journal just before setting my alarm.

Four hours later, I am stepping out into the cold, clear air. Bastien buckles the final strap on his life vest, and then he, too, walks through the hatch onto the back deck. He turns to me and whispers, "Welcome to the calm before the storm."

"By 'storm,' you mean—"

"The next thirty-eight days," he says, then laughs.

Jennie pulls her gaiter over her mouth and starts to kick chunks of ice through the holes in the bulwarks. Bleary eyed, I stamp my feet and clap my hands, then walk to the railing and hang over it, looking down. The sinuous roils our thrusters spin into the water's otherwise placid surface mix the sky's reflection into the sea, folding mauve into lavender.

The half-light not deep enough for stars.

Eventually, Jennie straightens, turns, and nods. Carmen, one of the MTs, holds his hand above his head and pinches his gloved fingers together: a signal to the crane operator to lift the submarine into the air. Ran's bright-yellow harness snaps taut as the tension in the cable increases, then the sub lurches free from her cradle. Her two-ton, torpedo-shaped body sways out of time with the ship and the sea. The large metal arch from which Ran is suspended looks like the gate to a Shinto shrine, marking the threshold between the sacred and the mundane. On this side of the metal superstructure: the *Palmer* with its library, desalination system, leftovers on the buffet, and fifty-seven human beings, some of whom are tucked into their bunks beneath heavy comforters. Beyond the crane: the Southern Ocean bathed in rose-colored light, the surface silky pewter and the edges of the bergs radiant red.

Filip, the long-haired, Swedish master's student, keeps Ran from batting against the crane by nudging it with a boat hook. The gate lowers toward the Amundsen; then the cable lets out and the submarine enters the water. Ran bobs in our gently churning wake. Ten meters, twenty meters, thirty meters away she drifts.

Nearly everyone from the TARSAN team stands on the stern to watch the submarine dive. Johan shoves his hands into his pockets and sighs.

Finally, Anna sends the signal to start the mission. The propeller spins, kicking up water, and the nose pushes down. But not enough. The propeller spins again. Just beyond, Burke Island rises out of the sea like the long low hump of a whale's back—white, like Ahab's obsession. Everything is warping: night is day, land appears mammalian. The normally clear boundaries between all kinds of matter blurring fast.

The buoyancy is off. The density of the water itself miscalculated. Not enough of the propeller sits below the surface, which means it lacks the necessary traction to nudge the load downward. The mood heaves from nervous anticipation to funereal disappointment. The second test launch declared a failure, and a Zodiac boat lowered overboard to catch the sputtering sub. Jennie drives. Filip and Johan clip a giant carabiner to the harness to drag Ran back to the ship. Before the papers can be written and the journals solicited, before the models can be improved, the data must be got. Gathered from a sea that does not release its secrets easily.

Skuas dive-bomb the inflatable as it draws closer. Anna shoos them away before slumping forward in her jet-black, subzero body armor. The skin below her eyes sunken and ashy, her lips downturned. The only person who can solve the problem with Ran is Anna. While I want to stick around and ask her about the difficulties and discipline that ground her work, I also know she requires rest, not needling. I walk out of the wind, hang up my hard hat by the door, return my float coat to the mudroom, and climb into my bunk.

For a long time, the sleep I think I need doesn't come.

Part Two | *Islands*

SETTING: *Water sky* (noun): In the deep south, when light hits the blue of the ocean and lofts back into the air a bruised color. This is no storm warning. Here, it indicates the opposite: safety, something to be sought, open water instead of ice. The ship sails all morning through this blue tone until it reaches some land.

After half a month at sea, the seeming permanence of stone shocks. Volcanic ridges rise out of the ocean like the teeth of an old bread knife. I count six islands, each no more than a kilometer long. At their edges, where rock meets the Amundsen, patches of snow persist, carried over from winter, when cold grips the water surrounding the Edwards Islands. I trace my eye along their low-slung crags, and the palette shifts from blanched bone to mauve, colored by guano from the thousands of Adélie penguins that call these outcroppings home. Just how long ago did the West Antarctic Ice Sheet withdraw, revealing the coarse, contoured ground beneath? This is what we are here to find out, which is why we have allotted a couple of precious science days to exploring these islands before we continue farther south to Thwaites.

"We'll be the first people ever in Earth's history to set foot on some of the smaller ones," Bastien told me the previous night. At the time, his declaration thrilled me. But now, looking out at what the retreating glacier revealed, I feel depleted by it too.

The *Palmer* drops no anchor but holds its heading through dynamic positioning, the thrusters and propellers programmed to keep us in this particular place. Becky joins me out on the bridge wings to watch the launch of the first Zodiac. The boat looks like a plaything, a little Lego ship, stupidly small in comparison to everything else. Its gray rubber hull skitters across the sea, then rounds a point. In the time it takes me to snap a photo, my fellow shipmates have disappeared. One blink and they're gone.

Just beyond the islands: a rumpled wall where the ice sheet meets the sea, its ragged cliffs bisected by brilliant veins of cobalt. A giant white dome rises up above it, bloodless and bright. The

Canisteo Peninsula looks nothing like what I would call land. What I see is endless and it is ice, its opalescent contours catching and refracting the watery, midday light.

Becky pulls her lavender headband down over her ears and asks how to describe the strange blue humming in the places where the ice appears pocked. "It's almost as if that blue carries you back in time," I say.

"Spoken like a poet," Becky responds. Then she adds, "But it's true, when we look at that color, we're actually journeying into the planet's past." Long before Canisteo Peninsula was so named (after the USS *Canisteo*, a tanker that provided fuel for Operation Highjump in 1947), long before these islands that rest between it and us were called Edwards (after the officer in charge of navigation on the Bellingshausen Sea Expedition of 1960), the snow that would form this glacier fell from the sky. The layers of flakes first compacted into firn—a sugary rind—that with time and weight crystalized into glacial ice, the little air pockets it contained flattened into long cylinders and tubes. This dense atomic structure eats all the colors in the visible spectrum except for blue—an ancient, ethereal blue. What we now gaze upon is fashioned from precipitation that dropped before the rise and fall of Rome, before Jesus or the Buddha were born, before the invention of the alphabet. Before sound became symbol.

Back in Alabama, Becky's three sons are likely at the soccer field, she tells me. The toddler is too young to play. L—, the oldest, recently lost two of his baby teeth. "My husband's struggling to stay on top of lunches and laundry and snacks and getting the kids to school," she says. "He's starting to burn out, so we're thinking of also getting a sitter."

While the pale sun struggles through the clouds, I wonder how to reconcile the details of our human lives with the fact of a glacier, the multitudes it holds. The oldest ice cores ever drilled originated in East Antarctica and contain a record of past climates ranging back eight hundred thousand years. They tell us that during that time, CO_2 levels have never reached as high as they are today. They tell us, if only slantwise, that human beings are altering glacial facts with our soccer vans and our diesel-fueled ships as large as the fields our children play on.

"Something in my gut says we shouldn't be here," I say. For the first time since setting sail, I blink back tears. They promptly freeze in place on my wind-burnt cheeks. I hadn't imagined how profoundly apart this place would feel—how it would appear gigantic and fully formed, an entity all its own, well beyond the limits of human understanding and resistant to whatever language I might try to pin on it.

"Becky," I say, lifting my voice over the endless dull roar of the engine. "Let me know if I can help you guys with sampling. It's the least I can do." I am aware, like everyone else onboard, of yet another irrefutable fact: in a few weeks, the ocean will freeze over, locking us out. There is a clock and it is ticking.

"Sure," Becky says, surprised. "I want to get a few cores under our belts, and then"—she claps her hands to encourage a little blood back into the tips of her fingers—"and then we'll see." She turns to look at me, to try to judge, I think, the quality of my offer, whether it comes from a genuine desire to help or is motivated by the journalistic impulse. She's not wrong to be skeptical. The writer in me doesn't ever fully turn off. Instead of trying to explain, I turn my attention back to the half-frozen sea.

A dozen or so Adélies wobble in a line across a nearby floe, finally flopping onto their bellies and launching into the ocean, flippers akimbo, feet outstretched.

CONVENTION SAYS THAT IN ORDER to tell a story about Antarctica, you need characters and you need plot. Most would specify, if asked, that humans must be the protagonists and play all the supporting roles. But after spending the day looking at an unfathomable sea that soon will transform into something solid, fused to a continent that carries traces of atmospheric history hundreds of thousands of years old, I'm starting to think that convention is absurd. Everything around us seems like so much more than backdrop.

In *Do Glaciers Listen?*, Canadian anthropologist Julie Cruikshank chronicles how three distinct worldviews—"Indigenous, Western, and scientific" as she terms them—understand the Saint Elias Mountains of northern Canada and Alaska. In some Tlingit and Athapaskan stories, glaciers take action and respond to their surroundings. They generate transformation, giving birth to seasons like summer and winter; at other times they swallow humans whole as a warning that others should act with greater humility. "Elders who know them well describe them as both animate (endowed with life) and as animating (giving life to) landscapes they inhabit," writes Cruikshank. Even though these Indigenous people live about as far away from Antarctica as humanly possible, I suspect they know something I don't. That the millennia they have spent alongside the ice has made them keenly attentive to its movements and meanings, which is why *Do Glaciers Listen?* is one of a handful of books I've brought along.

One afternoon, while interviewing Rob in his cabin, I ask whether we might think of Antarctica as not just shaped by human actions but also as shaping us. For a long time, he says nothing, and I fear he is trying to figure out a nice way to dismiss the idea as humanistic sorcery.

"I don't think I'm going out on a limb by saying that the exceptional stability of the Holocene plays a role in the rise of our particular kind of human civilization," he finally responds. "That idea has been around for a while, at least since the midnineties. Paleo records show us that over the past six thousand years, sea levels have barely changed at all. But if you zoom out just a little bit, you realize that stability is exceptional. Between eighteen thousand and six thousand years ago, the earth was losing ice very fast, and sea levels were rising, on average, ten meters per millennium."

"Three feet per century?" I say, converting between metric and standard.

"Some say that three feet of rise by 2100 is a ridiculous exaggeration, but the long-term record says that's pretty average, says sea level steadiness is the exception, not the rule," Rob responds. Then he leans back in his green upholstered chair. Were you to search for *Antarctic Explorer* in the encyclopedia, the illustration would look a heck of a lot like Rob. He's got the beard, the steely gaze, the quiet resignation. This is his twentieth voyage to the Southern Ocean; he's come so often he's almost lost count. But the more time I spend with him the more he surprises me, thinking and acting differently than the script I imagined for him.

Together we talk about how what happened in Antarctica—or, more precisely, what didn't happen: ice sheet withdrawal—influenced our history well before humans made direct physical contact with the last continent.

"Antarctica's stability might have been related to the importance of port cities, international marine trade, and the spread of settler colonialism," I venture. "If sea levels were rising three feet a century while Europeans were 'discovering' the Americas, things might have worked out differently." Were they not so eager to steal land and enslave people to work that pillaged earth, then we might live in a present where the cost of combusting crude, among other things, would be higher. The distance between us and a livable future not so vast. The longer we talk, the clearer it becomes that the entire cryosphere—the forty-six thousand glaciers that weave among the world's highest peaks in the Himalayas and among the mighty basalt bowls of Patagonia, all the parts of the earth where water is locked in cold suspension—dictates many aspects of our lives. That the geologist sitting before me agrees is a revelation, freeing me somewhat from the uncomfortable feeling that I'm just taking an Indigenous perspective from the other pole and using it to project what I wish to see onto the ice.

The Icelandic writer Andri Snær Magnason takes up this subject of the stabilizing, cradling—even nurturing—effects of the ice in his book *On Time and Water*, a meditation on the power and promise of living in conversation with glaciers. When he learns that both Nordic mythology and the Vedas, the oldest Hindu scripture, suggest that a sacred cow (in some instances fashioned from hoarfrost) gave birth to the world, he writes, "Suddenly, everything seemed to click, all at once. . . . A cow made from rime is the perfect metaphor for a glacier. A glacial river contains no ordinary water at all: it flows milky white with dissolved minerals." In Iceland, as in the Indus Valley, nearly everyone depends upon the seasonal pulse of meltwater pouring out of the high peaks. That origin stories from both locales propose a symbolic link between glaciers and mothering, between their ability to nourish and the way they create communities, doesn't surprise me.

"The public only becomes aware of how dynamic the earth is when change occurs on a time scale that matches our own," Rob says. "When it's something they can see. Volcanoes erupt, tsunamis hit. And for a little while, people understand the power of geologic forces."

I squint at an iceberg out the porthole above Rob's right shoulder and wonder whether a glacier loosed her into the rising sea as punishment for four centuries of attempting to bend the planet to a particular set of human aims—or whether, perhaps, the berg is an invitation, the surprising rate at which it and so many others have been released by Antarctica, a demand that we acknowledge our interconnectedness. Retribution or reconciliation? It occurs to me then that how you interpret the glacier's movements likely depends on where you stand and what you might gain or lose in a remade world.

JOEE:

I went to sleep the night before having been told that,
when you wake up, we'll have been at the islands for
three hours. Everyone will be ashore, and your job
will be picking them up later in the day. But, as is very
frequently the case, the plan didn't go according to plan.
Some people were on the islands, but others had yet to
make it out. It was, like, eleven-thirty in the morning,

and the reality is that the outgoing shift [of MTs] had just worked twelve hours straight, and the oncoming shift was about to work twelve more. I was like, *Guys, I need twenty minutes to wrap my head around what I'm about to do.*

CINDY:

When we were loading the Zodiac, it was bouncing and taking in a lot of water. The rope had to be let out as the Zodiac went down, then I had to pull it taut to keep the boat from slamming against the *Palmer* as the waves rose up. I just kept letting out line and pulling it in, trying to keep the rope tight. But as the Zodiac bounced, it created waves too. That really sent it flying.

JULIAN:

I was on the helo deck, freezing my butt off, trying to grill some steaks. I marinated 'em in pepper and garlic salt, sautéed 'em in the kitchen, then took 'em out to the pit, where I threw some olive oil, some soy sauce, some Worcestershire sauce, and a little bit of sugar on 'em to break down the acid. I couldn't feel my hands. I had to run inside and run some hot water and put my hands under there for a few minutes. Then go back out and flip the steaks.

JOEE:

The conditions were bad. We were seeing short, steep seas. There was lots of open water in the direction of the wind, lots of fetch. And it was very cold. Lots of spray. Periodic bands of brash ice. Whitecaps.

GUI:

Eventually, we got the boats out. We were heading to the islands. We knew a few places where there would be some seals that we saw from the ship. As we got closer, we were hoping to see females ready to be tagged.

JOEE:

We were getting our asses beat up on the Zodiac. Getting all tossed around.

GEORGE:

The MTs always need some extra hands, and I've been here long enough to be able to do many different things. So I was invited to take a ride. It was my "reward" to be on the inflatable, to get to really see what's around.

MEGHAN:

We went around the island one way. Then we went around the other way. It was surrounded in shoals and sunken ridges where the waves broke. Some sides were covered in ice cliffs. The first island was a crappy island to land on. The second island had too many penguins; we can't get too close to their rookeries. Finally, at the third island, George jumped in the ocean in his dry suit and walked us in.

GEORGE:

I was helpful. And that really is what my job's about, being helpful. Everybody likes that part, though, that's why we're here.

BASTIEN:

The beachmaster [a male elephant seal], he was—according to Lars, who's the expert authority on the matter—over two tons, probably approaching two and half tons. That's as big as a big car.

MARK:

It was my first time going near something like that. Baptism by fire. Lars was telling us, *We just do this, then that, and they'll be on their own and it will be nice and easy.*

[*Pauses*] Yeah, it was interesting. There were massive, just massive, males in there. And our job was to wave a stick around in front of them to get them to move out of the way, so we could tranquilize and tag the females. I think the biggest was about four meters, five meters long and two and a half tons. If you cut the legs off a big cow, he was bigger than that.

BASTIEN:

If you cut the legs off a cow, that cow would fit in the beachmaster's stomach, that's how big he was. He would arch his back and rock his weight back and forth. The front end is the pokey bit that could bite you or throw you across the beach. The tail is a pair of massive flippers that are ninety-five percent muscle. If those whack you, they can send you flying. He prevented us from getting to the females, of which there were four at the start.

KELLY:

The seal team went out first, then we found an island that suited us. We got all our gear together and moved it up out of the wave zone, then we put on new gloves, got warm, and traipsed up the hill. Finally, our team was also off and running.

MEGHAN:

We got to our spot and pulled all our field gear over to a nice little windbreak.

JOEE:

There's this great book called *Folklore and the Sea* that's all about how sayings like *Mackerel skies and mares' tails make tall ships carry short sails* have a truth to them, because they really are references to centuries of lived experience. People from one place could predict the weather three or four days out, because they were like, *Oh, this particular flower is wilting today.* Or, *There's a certain ring around the moon*, or, *The wind shifted, and I could smell apple blossoms.* There are all of these really subtle, delicate signals that are embedded in human memory, and they connect us to the environment. But Antarctica is the only place on earth where there is none of that—like, none of that at all. There are no Indigenous people who can say, *When the wind shifts and comes around from the north, you know it's going to blow like hell for three days.* And without that knowledge, that deep institutional knowledge, when the barometer drops, you really don't know what's going to happen next.

MARK:

On the second beach, we found a seal that was a solid four meters away from the water. That gave us enough space to place our gear and put people in front of her to be able to discourage the seal from going in the ocean.

LARS:

I had the dart in one hand, the blowpipe in the other. We were walking up to the animal and pressurizing the dart—so about ninety seconds from knocking her out—when the radio call came in. The captain said, *We've got winds up to forty knots. Cut mission.*

VICTORIA:

We heard them say over the radio that they were going to have to get the seal guys first. Then we packed up and just stood there making fun of the penguins while we waited for the Zodiac to come get us. There was one baby leaning against a rock, giving us this look like, *How did I end up in Antarctica? I didn't choose this life; this life chose me.*

LARS:

I'm not disappointed, per se. I know that this can happen. I'm a bit sad. It could have been nice and easy. But, you know, it was the perfect seal, so obviously we had to abort. There will be more. The real issue is that the bad weather is cutting into time. I know there are seals. I know where they are. We can tag them. I just hope they'll be there when we come back, but we might not have time. If we stayed for three weeks around the islands, I would be a happy bunny. But I think everyone, myself included, we really want to get to Thwaites.

UP ON THE BRIDGE, I pull a reference guide to the Beaufort scale off the shelf. First used by the HMS *Beagle* during Charles Darwin's circumnavigation of the world, it remains one of the most popular systems for measuring the wind while at sea. The idea is simple enough: water does specific things when wind blows across it at a specific speed. And so the scale, which ranges from zero to twelve, uses basic observations of the former to provide a reliable calculation of the latter. For instance, large wavelets with breaking crests register as a three on the Beaufort scale, which means that the wind is moving at seven to ten knots.

I alternate my gaze between the window and the photos in the guide. Outside, the crest of each wave crumbles into long white fingers that drag across the platinum sea. Its marbled surface undulating to the point of pulling apart.

"What's that, a seven or an eight on the scale?"

"Try a nine or a ten," Rick says. "Flip forward a couple of pages."

"Scarlet Begonias" tinkles through the speakers, fills the room with its warmth.

Beaufort Force: 10. Wind Speed: 48–55 knots. Very high waves with long overhanging crests; the resulting foam, in great patches, is blown in dense white streaks along the wind direction.

The ship gives a deep, low bellow.

"Fifty knots. What's that in miles?" I ask.

"We're easily getting gusts up to sixty," Rick says. Just a little faster, and we would be in the middle of a hurricane. The survival bags that each team must take to the islands suddenly make more sense. Waves peel across the water, spindrift pulling from the surface lacy streaks of white. The *Palmer* steams away from the islands and toward the relative calm of Ferrero Bay, a small, ice-free inlet just south of the Canisteo Peninsula and a full day's sail from Thwaites.

The storm brings a strange calm to the inside of the ship; without data to be gathered, the dinner-table talk meanders and somehow turns to labor. The excruciating hours, sometimes days, it takes to birth a child. I hear about a husband who nearly fainted and the woman who drove herself to the hospital. The curving country roads and the contractions, how it seemed the only way out was through the pain. I expected these scientists to be even more anxious about procreation than I am, but I have learned that most of the adults onboard have at least one kid back home. Becky, Victoria, and Rob all have three.

"I'm hoping to have a family, but I had to postpone trying to get pregnant so I could come here," I confide, savoring the ease with which our conversation slips between the two topics that fascinate me most. "When I started this project, I worried that the more I knew about Thwaites, the less I would want a child, but that hasn't happened yet." I mention an article recently published by *Proceedings of the National Academy of Sciences* suggesting that, even if we quickly shift to a worldwide "one child" policy, we will still have five to ten billion people living on the planet in 2100. Its underlying argument—that rapid transition away from fossil fuels, not fewer pregnancies, is what is needed—gives me some solace. "Population control just takes too long," I add as I push peas around my plate.

Rob excuses himself, brings his dirty tray to the dishwashing sink, and leaves. Victoria and Becky start giving me breastfeeding tips. Kelly picks up her bowl, stirs her rice into her beans, and takes a bite.

"How many times do you think giving birth has come up in the *Palmer*'s mess?" I ask.

"Maybe never," Victoria offers, nodding at the crew's table, where Captain Brandon and Kiel, the third engineer, wordlessly plow into their plates of soul food.

Together we share a secret, furtive laugh. It occurs to me then to formally ask each of my shipmates to recall the stories of their own births. The question feels transgressive on a boat destined for a continent where women were so recently all but banned. Beneath the allure of taboo, I also sense a unifying potential. Each of us was born, after all. Our presence on the ship impossible without that creative act.

Becky screws up her big brown eyebrows, leans over her empty plate, and says, "I've never been away from my children this long, and some people criticize me for it, but I want my kids to know that their mom made decisions to be part of something larger than herself." For over a week, the internet has been functionally out of commission. If Becky wants to speak to her sons, she has to use the Iridium satellite phone in the computer lab and limit her conversation to ten minutes. "They're surviving," she says.

In "Objectivity or Heroism?" historian Naomi Oreskes argues that scientific inquiry has traditionally been valued in proportion to what a person risks to accomplish their research. For instance: Walter Reed is celebrated not just for discovering that mosquitos transmit yellow fever, but for supposedly exposing himself to the disease. We remember Robert Falcon Scott for his retrieval of the *Glossopteris indica* fossil, a 250-million-year-old "fern-tree," which proved that Antarctica was once part of an ancient supercontinent called Gondwana. That he died while lugging back that old rock only deepens our admiration for the discovery.

But instead of being celebrated for what they are willing to give up, women working at the poles are often considered burdensome: not only does their presence require "additional" support on the ice, it also means they are withholding support they are expected to provide back home. When I prepared to deploy to Antarctica, two of the most common questions I got were "Is your husband going with you?" and "Are you leaving kids behind?" Each time it felt like an attack. I imagine Rob on the receiving end of a similar inquiry and start to laugh—that's how absurd the question sounds in reverse.

The *Palmer* lets out a mournful groan as a chunk of ice grinds down the length of the hull. On the partition behind Kelly rests a bread box–sized aluminum replica of the ship none of us have left in nearly a month. Just to its right hangs a cartoon of the 1998 sea-ice cruise, depicting men with jaunty facial hair as they drill ice cores, wield handsaws, and smoke cigars. Every time I notice it, I want to roll my eyes, first at the tired display of Antarctic masculinity and then at myself for expecting otherwise.

Becky gets up to give her kids a call. I stand and carry my bowl over to the dishwashing sink, where the same man who cooked my dinner will soon scrub my dirty plate.

"Thank you, Jack," I say, scraping my leftovers into the trash.

"Y'all are so good at appreciating us," he responds, his braces flashing in the fluorescent light.

"One of these days, I'd be happy to help you guys," I say. "My job is to get to know—like, really know—what goes on around here."

"I think you need a certain kind of certificate to work in the kitchen, but I'll check," Jack says. It is not a coincidence that he and Julian are the only two Black people onboard and that they are both employed in the galley. Nor is it a coincidence that all the able-bodied sailors hail from the Philippines. As I watch Jack push the plate into the soapy water, I contemplate a whole different lineage of stories missing from the Antarctic archive: those of the crew members who tend the ship and her passengers. Theirs is a labor of maintenance, not of discovery. It will not be celebrated in the pages of *Nature* nor *The New York Times*. The basic things that our lives depend upon—cooking, coiling ropes, clearing decks, navigating—are all so underappreciated that they go without mention in most accounts of polar science, especially today.

Writers and philosophers from Herman Melville to Michel

Foucault have imagined ships as spaces apart, where isolation generates the conditions for a redistribution of power. But despite our sharing many of the spectacles of Antarctic travel with one another—watching the first icebergs arrive on the horizon or a storm blow past the windows—the roles each of us play are not valued equally, nor are they equally accessible. Those who come from affluent or privileged backgrounds are more likely to work in positions that generate the kinds of accolades that in turn attract more of the very same resources that led to those opportunities in the first place.

After depositing my utensils in the dishwashing sink and tossing my paper cup, I walk down the hallway, past the computer lab, where Becky swivels on the stool by the phone. She laughs, tells her husband that if he's stressed, he ought to take the pressure off. "Go to McDonald's, it won't kill them," she says. I keep walking, past Cindy, organizing glass vials in the Dry Lab; past the door that leads to the engine room, where Kiel tends the machines that propel us ever closer to Thwaites; down to the end of the hall, where I turn the handle on the big green door that opens onto the stairwell to the upper decks. The same stairwell Fernando mops every morning, first thing, with attention as constant as it is overlooked.

KELLY:

You know what you need to do this work and have a family? Support. In the past, we've had my ex's mom or my mom come to help him while I'm away. His family is from Portugal, so when L— was younger, we even had an au pair once. She spoke Portuguese and cooked Portuguese food, so that was really good. Unusual and good. Every time I leave, the solution we come up with is a little different.

It's hard, but I'm also really proud of her. She's so strong and adaptable. She wasn't super psyched that I was going on this cruise. It *is* a bit longer. And she's a bit older—just turned nine—so her needs have changed quite a lot. When L— was two and a half, I went away, but she doesn't remember it. She will, however, remember this.

On the other hand, she's super excited by what I do. She loves science and she loves the outdoors, so I don't know. It's really hard to balance, but the more people do it, the more people understand it. It's becoming more and more normal. L— stayed with her dad this trip. This is the first time he has had her the whole time, and I think it's probably been good for both of them.

Once I'm back, I'm going to be off from work for a few weeks. L— has time off for the Easter holiday. She's counting down the days until I return. We already have plans for our first weekend together. We're going to cycle into town, then we get a cookie to reward ourselves for cycling into town. Then she wants to go see *How to Train Your Dragon 3*, which is just out at the movies. We'll have lunch out at an Italian café that she and I have discovered and made our own.

When we last spoke, she said, *Can we have pancakes in the morning?* I was like, *Absolutely*. Then she was like, *And can we make Rice Krispies cakes in the afternoon?* And I was like, *Well, sure—it's going to be a really sugary day, but whatever*. I'm used to it from being on the ship. There will be a bedding-in time again, and right now that sounds really good, necessary. The main thing is that we're going to do a lot of what we normally do, together.

When I return from Antarctica, Felipe and I immediately book an appointment with my sister-in-law's ob-gyn to talk about what we need to do to start trying. He schedules an appointment to draw blood, then he asks if I am taking prenatal vitamins.

"Yes," I say. "On and off for over a year."

Finally, he tells us to wait three months for the potential risk of Zika infection to clear from Felipe's sperm. During that time a thin curtain of rain falls continuously over Bogotá, turning its streets gray and conspiratorial. And yet, on the morning we return to the clinic to get my IUD removed, the sun buzzes brilliantly above. Our little red Uber sputters through traffic, past mom-and-pop tile factories, people pouring ají on empanadas, and kids carrying backpacks on their way to school. On the radio, Chico Buarque sings, obsessively, "O que será, que será."

There is no waiting room at the clinic, just six hard-backed chairs tucked into the hallway's far corner. At this point, we've postponed our plans to conceive for exactly one year. I take Felipe's hand in mine, then lay it flat on my thigh. It's hard to know how to feel. I count the succulents in the window box. Fear and excitement chase each other in my gut. What if I'm not able to carry a child? What if I am? Have we waited too long? And beyond all that: What kind of future will this being inhabit, should it choose to come at all?

"You know, I was born here," Felipe says absentmindedly. "Thirty-nine years ago."

A couple walks by, clutching a newborn wrapped up in a thick fleece hospital blanket. The secretary invites us in; the doctor flips through my bloodwork and confirms that everything looks good.

"Are you ready?" he asks, his English slow and deliberate.

"Yes," I say. "Yes."

Then he hands me a thin paper gown and tells me to wrap my body from the waist down.

LINDSEY:

When you meet someone who is one of thirteen kids—you know, that used to be necessary in terms of farmwork, but now the idea just makes me panic. One, I think: that's going to be my number. No more than that. It's such a strange correlation, that this is where, for me, the climate-change message manifests. I don't think I could or would want to have more, 'cause I've seen the impact we're having on Antarctica.

PETER:

I mean, I know climate change is bad, but to allow it to influence your life to such a degree, that I don't understand. Should I or should I not take a long-haul flight? That's a reasonable question to consider in the context of climate change. Should I or should I not have kids? That's something else entirely. You're allowed to have two children, because then you're keeping the number of people on the planet the same. If you want three, I suppose then we can start arguing about climate change.

BASTIEN:

My work on Thwaites feeds into a general unease that I carry around about the future. There's this really basic question: Is it a sensible choice to have kids at this point? I'm afraid of messed-up politics, the rise of what they call "populism," reactionary movements. Then there is, you know, food scarcity, water scarcity, World War III. That could happen easily in our lifetimes, and it would be fought over diminishing resources.

GUI:

There's the personal time scale and the humanity time scale. When I think about my son, I'm thinking about one hundred years from now, and that's about it. And I'm concerned with things like his happiness, his kids' happiness. I think I'm very selfish in that. When I say *selfish*, I mean, I just hope that the world is hanging on for at least two more generations and that the people I love are safe. But when I think about the humanity scale, I'm concerned about our survival. And not just us: all beings. One thousand years from now, if humanity is still around, people will look back and say, *That was the generation that changed everything. The world was collapsing, and they did something.* I want to be part of that legacy. But it's not a given.

ALEKSANDRA:

I do not have children, and I'm not sure if I want them. Climate change figures into it, because it's hard to have children right now. It's hard to educate them. People have changed, they don't care about things. When I was a kid and I had a candy wrapper or something, I had to carry this paper around with me until I found the trash can. And now I just see people throwing the same thing on the street and thinking someone else will clean it.

MARK:

I can't say that having a family hasn't crossed my mind. Let's not tell my partner quite yet. I'm thirty-five. I certainly think about climate change in terms of family size. You know: How many is a good number? What sort of impact are you going to have? I think all the impacts we're seeing now, they will see more of in the next generation. It's mainly just a size issue for me. The thing is to balance out our ideals against social pressures and practical concerns. We're lucky to be able to make that choice.

JOEE:

I decided a long time ago that I didn't want kids. I had a ton of responsibility for my siblings when I was young, and I was like, *I've changed all the diapers I want to change in my life.* So working down here hasn't impacted my desire to have children, but it makes me feel a tremendous amount of anxiety for the children that are in my life and that I care so much about. I balance that feeling out by trying to offer them inspiration. I love my position as the town auntie. My friends often ask me to talk to their kids about being a woman in Antarctica, and I'm like, *Yes! I will absolutely do that.* I don't like to be the center of attention, but I realize that I can provide something these young girls can circle around and be engaged in, a way of life that has this kind of awareness in it—of Antarctica and the ice—and that helps with the anxiety.

COME MORNING, THE *PALMER* IS encased in a thick layer of ice. A strange, anxious atmosphere permeates the ship. The storm has passed and the weather is fine, which means that the seal crew and those heading to the islands to gather organic material are finally in a good position to get a lot done. And yet opting to do that means staying here and further delaying our arrival at Thwaites. The night before, I offered to help Scott and Meghan search for ancient penguin bones. "I'd come not as a journalist," I assured them, "but as a field assistant." At this point, I'd do anything to get off the boat, even if just for a few hours.

Unlike Becky and Jack, Scott didn't balk.

"You're hired," he said. "It's not that hard. That's why my advisor sent me to do it for her."

So now, I fill a bowl with Ariel's famous *lugaw*, a dense rice porridge from the Philippines, and stack four pieces of bacon and one fried egg on top. Fill another bowl with sliced kiwi. "Amazing these haven't spoiled yet," I say to Lars as I sit down next to him.

We high-five, then eat what might be our last warm meal for a very long time. Like a kid on the first day of school, I try and fail to picture what the next twelve hours will hold.

After breakfast, I carefully arrange my expedition clothes on the worn white countertop in the aft Dry Lab. For the bottom: two sets of long underwear, cross-country ski pants, lined Carhartt overalls, and a fisherman's bib. For the top: the same base layers, plus a sweater, soft-shell jacket, windbreaker, raincoat, balaclava, wool hat, waterproof gloves, and float coat. Into my waterproof backpack I shove two pairs of backup wool socks, as well as extra pants, jacket, sweater, hand warmers, hat, hiking boots, and two pairs of gloves. The first rule of Antarctic fieldwork: always bring a change of clothes.

"The windchill is minus thirty," Lindsey says, holding out a blue plastic bin full of snacks. "Be prepared to eat to stay warm." I rummage through her selection and grab a Twix, a Snickers, and three Nature Valley granola bars, which I toss into a Ziploc along with the three pigs in a blanket I lifted from the mess and one of my last bags of gummy bears. Meghan takes a bunch of chocolate bars and places them on the countertop alongside a maple syrup–covered pancake and three slices of bacon. "I'm not hungry first thing in the morning because of my thyroid medication," she says. "But I know I need to eat something." She removes her extra GPS batteries from their charging unit and places them and three pairs of gloves in her pack. Then she picks up the pancake and takes a tentative bite.

Each country is responsible for outfitting their scientists with appropriate outerwear. The gear from their government makes Anna and the rest of the Swedes resemble a battalion of Starship Troopers, sleek and battle ready, while the Brits all sport the kind of orange jumpsuits you might find in a car mechanic's shop. The clothes the NSF issued, on the other hand, make us look like a bunch of scrappy fisherfolk or a flock of winter-ready rubber duckies. When Tasha waddles over, her rain pants squeaking, I laugh.

"Get your gear on," she tells me with a smirk. She's younger than me and sinewy, a former sailor in the Navy who enjoys giving orders. In almost any other situation, I'd rebel, but I know I have no idea what I'm doing. After putting on all my layers, I have to lengthen the strap on my backpack just to sling it over my shoulder. We walk together down to the Wet Lab, which is crowded, stinky with gear and underslept scientists. Three pairs of damp, penguin poo–smeared Carhartts hang in front of the fan, discarded by those who went out the previous day. There are a bunch of purple surgical gloves scattered around the sink. In the center of the room sits a pile of pickaxes and shovels, next to half a dozen pony-keg-size spools of rope.

"Fuck," I sputter, banging my knee on one of them as I try to step out of Lindsey's way.

Scott shoves three tubes of Pringles into his pack. "You know, those are some of the most difficult things in the world to recycle," Lars says to no one in particular. "Because of their plastic lining and their metal base."

Outside, a frigid funnel of air thrums around the ship. Jermaine turns the crane on and the Zodiac lifts off the helo deck. It swings beyond the bulwarks, then floats down to the sea's heaving surface. Machine hum drowns out the MTs' discussion of where to attach the Jacob's ladder. Fernando carries the bundle of wood and rope to the stern and screws it into place. Jennie takes one look and changes her mind, gesturing toward the starboard side.

"It's a little more in the lee," Lindsey yells above the noise.

The scientists grumble and scatter. Lars stomps over to the railing, uncaps his binoculars, and starts searching the rose-colored humps of land for fully molted females. Today is the second-to-last day set aside for island work, and every minute that passes is potential data lost. There are so many things that can go wrong. But I am learning that "wrong" isn't exactly the right word—for every proposed bit of fieldwork, there are four or five backup plans that the scientists cycle through with surprising nonchalance. Still, the storm cost us a day. Although it might not sound like much, for Lars it means that some seals he could have tagged will have already left.

"Classic example of hurry up and wait," Bastien says, pushing his ski goggles onto his forehead. His mustache mottled with frozen breath. Fernando removes the ladder, rolls it up, and resecures it to the starboard side. My legs sweat.

"Bridge, Zodiac one is in the water," says Jennie, her husky voice bouncing out of Lindsey's radio.

"Roger. Zodiac one in the water. Copy that."

"Seal team, you're up," Lindsey says, waving the men over to the gap in the *Palmer*'s railing. Gui flashes the hang-ten sign and walks to the ladder. He turns his back to the sea, grabs hold of the rope, and begins lowering his body over the side of the boat. Then he's gone. Bastien follows. Then Mark, Tasha, and finally Lars. George hands toolboxes containing the seal-sedation kits down to the crew. They too disappear.

"Bridge, back deck. Are we clear to launch the Zodiac?"

"Zodiac, you are clear for launch," says Luke. I look up and see his profile, with his blue North Carolina cap, in one of the bridge's back windows.

The engine kicks up a key and the inflatable zips into view. Sunlight blisters the surface of the sea. I watch the little boat as it disappears behind the *Palmer*'s bow.

"Bridge, Zodiac one away with six passengers, headed for the western side of the Schaefer Islands," Jennie's disembodied voice crackles. Blood suddenly thrums against my stomach walls as I head inside to grab my bag.

Kelly comes in not thirty seconds later. "They're on their way back," she says.

"No way," Meghan responds.

"I guess there's a hole in the boat, ice punctured from the storm, maybe."

Scott places the shovel he just picked up back on the floor. Then he touches his hand to his red beard and shrugs.

"We have to get that patched up before we can send you all out—that or exchange the boats," Lindsey says. No one wants a leaky dingy in the Southern Ocean. Still, some part of me thinks, *Oh, can't you just get them on the damn islands, then fix the hole?* That's the part that knows the failures are heaping up, knows the scientists desperately need to get work done. The chance won't come again. Then I remember the whitecaps and hurricane winds from the day before.

In the Dry Lab, I step out of my plastic pants and place them back on the countertop. Lars walks in, but this time we do not high-five. "You've got to laugh to keep from crying," I offer instead.

"Maybe toward the end we'll have a whole week free, and we can come back," Lars says. "If only I could get a sugar daddy to extend the cruise." We snicker together at the absurdity of the idea, and my attempt to cheer my shipmate turns into him cheering me. An hour later, we're back in the mess, filling our stomachs again. It makes me nauseous to layer fried chicken on top of rice porridge and bacon, but I have no choice. If we get out after the midday shift change, we won't be back before midnight.

ALI:

Persistent uncertainty is something you get used to on a cruise. Even with the best-laid plans—that you have down to the minute—there are changes to the schedule, like really significant changes. It's just continuous: modify, modify, modify.

ROB:

You accept that Antarctic research won't always go smoothly. It's very common that something happens to disrupt your beautifully planned schedule. You may have noticed we don't plan in great detail what will happen many days out, because that would inevitably get knocked off course. You would waste a lot of time planning and replanning and replanning. We focus on what will happen in the next twelve hours, and how that feeds into the twelve hours after that.

ALI:

You have to learn to be flexible, incredibly flexible. Life on the ship is as much about the science as it is about getting used to everything changing from day to day, from hour to hour. I don't know, maybe that makes you more resilient to change, like in the larger sense.

KELLY:

You have to be resilient—physically, emotionally, intellectually—to constant change but also to monotony, because when you do finally start working, you're doing the same thing day in, day out, for months on end. Patience is what you need more than anything else, more than toughness or bravery or whatever else they say.

ALI:

There's a general, overarching plan, but the way it all shakes out at the end of the day is, well—I suspect it's a bit like painting a picture, like the work of an artist. You might start with an idea in your head, and by the end of the day you hope to have a good-looking painting, but it might not be exactly how you expected it to look, and the route you took to arrive at it might be different from what you imagined when you started out.

Just before we move from Bogotá back to Providence, I believe, briefly, that I am pregnant. When my husband lays his arm across my belly, I move it, not wanting to put extra pressure on the blastocyst that might be burrowing deep into my uterine walls. That night, at a friend's birthday party, I refuse the wine I am offered, instead ask for seltzer. His wife raises an eyebrow when she hands me the glass. I shrug and try not to smile. Ever since my return from the ice, I've been relishing what might be the last months when my time is wholly my own. I write in the mornings and take long walks in the afternoon.

The next day, on my way back to our little apartment in La Soledad, I stop at the corner market and purchase an at-home pregnancy test.

In the morning, I pee on the stick, then place it on a shelf in the bathroom, convinced that I will soon see two pink lines. To kill time, I walk around our bedroom, trying for calm. My hands shake. Back in the bathroom, I lift the plastic sheath.

One line.

Tilt my head sideways, hold the test to the light. Not even anything faint. Still, some part of me hopes. But four hours later, a rust-red coin appears in my underpants. In the bathroom, with my Jockeys around my ankles and the cold antique tiles beneath my bare feet, I become momentarily distressed. Think: *Maybe praying will help, maybe crying will help. Maybe more vitamins or even healthier food will help.* Then I start to fear that my reproductive system is broken, that hitting puberty early means I already used up my limited lifetime supply of eggs. Am I allowed to mourn a thing that never was? Who will sit beside me in this place of grief? It feels self-indulgent—and maybe it is—so I try to tamp down my disappointment, to brush it off and begin again. I unwrap a tampon I never wanted to use, insert it, and walk back out into the apartment.

Tick, *tick*, *tick*, goes the clock.

After I tell Felipe that we will have to try again, I get caught up in that word, *try*. As a verb, to *try* means working toward something without guaranteed results. *Perhaps that is what parenthood is*, I think, *a whole lot of trying without the promise of anything in return*. I give up my beloved habit of social smoking and limit my drinks to one. When Felipe offers to lift my bike up onto its mount, I let him. We start using an ovulation tracker and also a special kind of lubrication called Pre-Seed, which apparently "creates an optimal environment for sperm on their journey." The longer we try, the more I think about another definition of the word, as in, "those were trying times"—or "this is trying my endurance." Becky told me on the boat that many people she knows got pregnant only after they stopped trying. But nothing in my life has worked like that before.

SCOTT:

I was born in New Britain, Connecticut—maybe Hartford?—well, somewhere in Connecticut. I know that much for sure. I have no idea what my dad was doing. There are no photos. I wouldn't be surprised if maybe he was working. Which isn't to be insulting—the guy just works: that's all he's ever known and all he's ever done.

He came from a family with twelve brothers and four sisters. I think popping out kids was just another day on the farm. They had a potato farm, my grandparents. My dad and his brothers all worked on it in northern Maine. My dad was a tree cutter and then a construction worker. My mother was a real estate agent and raised three sons. She also raised horses back then. Now she does stuff in the dog world, breeds dogs and shows them. She absolutely loves animals, she always has. She switched from horses because they're so expensive. You often lose money with horses, whereas with dogs there's less overhead. If you get into the show world and have a good pedigree, you can even make some money with dogs.

But my birth? Really, I know next to nothing. I know that I was the third child, so my mother had been through a couple. I know that I'm the only one that was born with a cleft lip. When I came out, I probably looked a little ragged. It's something a baby develops in the womb. It can be a byproduct of smoking while you're pregnant, but there are lots of reasons why a child develops one. My mother had quit several years before.

I have two different-sized nostrils, if you can see that through my beard. [*He tilts his chin back.*] They took cartilage from my ear and reconstructed my lip. I had to have three or four surgeries when I was young. I imagine that kind of dominated the day.

I AM ALREADY READY WHEN Lindsey tells our group to stand by, and, much to my surprise, soon thereafter I feel the vague outline of the rope ladder in the palms of my double-gloved hands. With my back to the sea, I bend my left knee and stretch out my right leg, searching for the next rung. It is farther away than I expect, my stomach temporarily uneasy as I lower my body down. Finally, my toe touches something solid. I set both feet on the little wooden plank and take a long breath before continuing to descend.

"Go now," Joee barks from below when I reach the end of the ladder. She revs the engine to ram the Zodiac snug against the *Palmer*'s hull. I release the rope and fall backward into the inflatable, butt first.

"Good. Ready for gear?"

"Yes," I scream over the thudding of hard rubber against steel. Nod wildly.

Scott dangles a shovel over the side. I nod again, and he lets it fall into my outstretched arms. Two pickaxes, two more shovels, and two emergency survival kits all come down in quick succession. At last, he joins me, Meghan, James, Kelly, George, and Joee in the pitching dingy.

"Everything secure?" Joee asks.

George adjusts lines. "Yes," someone finally says, and the Zodiac skitters away from the ship, bouncing against the chop. My grip on the rope running along the gunwale tightens. A wave breaks over the bow, its frigid spume smacking my face. Fear pitching my heart rate higher. But when I clear the water from my ski goggles and look to my left, I see that Scott got drenched too. Meghan laughs at our misfortune. Then I am laughing along with her, gelid water dripping down the back of my neck and raw, relentless gratitude running through my veins. How sudden and visceral this contact with the ocean that has surrounded us, but which I have not, as of yet, had the opportunity to really touch.

Just over Meghan's left shoulder, the *Palmer* rises. For the first time in nearly a month, I behold our floating home in its entirety. The lifeboat, the winches and portholes, the rope ladder tossed over the side, the lights glittering on top of the A-frame. Part spaceship, part toy, the *Palmer* appears both sturdy and small silhouetted against the Amundsen and the scattered bergs. *This must be how astronauts feel during space walks*, I think, *when they gaze back at the vessel that has carried them among the stars.*

For fifteen minutes we zoom around the island, searching for a place to land. There are no obvious candidates, as all the ice-free areas are hemmed in by precipitous cliffs. After some discussion, we decide to try a cove with a giant ice wall on one side and a rocky point covered in hundreds of awkward Adélies on the other, their black and white coats glossy in the sun. Joee gets us close, then guns the engine to push the bow up against a pile of partly submerged stones at the cove's small crook. Everything starts to happen fast. Waves crash over the craggy granite. George grabs the line, hops into the surf, and, using his whole body, attempts to hold the Zodiac against the rocks while I work my way to the front of the boat and scramble onto the outcropping. My foot slips from the stones' icy skein. The sea sluices up around my ankles and retreats. Frantic, I press my palms into a waist-high ice embankment, shifting my weight to my hands. I hinge forward, swing my lower body up and around, and stumble to a stand, then step out of the way. Soon we are passing the shovels and other tools from person to person into the crevasse between the ice cliff and the rocks. When we are finished unloading, Joee backs the boat out of the breaking surf, to wait ten meters offshore.

I unfold the flap on my waterproof bag and dig out my extra jacket and balaclava. Exchange XtraTufs for hiking boots with added insulation. Trade wet gloves for dry and put hand warmers into the palm of each. I jot down a couple notes, then shove my audio recorder, pad, and pens into my jacket pocket and look up.

"How many of you have experience walking on glaciers?" Scott asks.

"I spent time in the Himalayas," says James, whose PhD research involves tracking glacial loss around the planet.

"I'm a New Englander," I add.

"Good. All you need to know is to walk perpendicular to the slope and dig in those toes—heels if you're descending. This isn't so bad, but you don't want to slip and go tumbling off an ice cliff into the ocean. The field site is only about half a mile from here, just on the backside of that big hump." He gestures to the snow-stilled headland rising above us.

I don't want the scientists to worry about me falling off ice cliffs or otherwise slowing their work down. "All I want to do is help," I say to Kelly, who is standing next to me, as I pick up a pickax and put on my backpack.

"Right," she responds, more sharply than I anticipated. Then she looks away. One of the mission's goals is to raise public awareness around Thwaites, but no one asked the scientists if they wanted to be round-the-clock climate communicators during their two-month Antarctic deployment. I have tried to be transparent when I am recording conversations and tried to step away if my presence seems burdensome. But it hadn't occurred to me until now that my being on the islands might undermine the liberation promised by a day away from the ship. I awkwardly fall in line behind Kelly as the five of us start our ascent.

The thin trail we follow is not the work of humans, but of thousands of penguin feet. There are no footprints, just snow tamped into ice. Despite the lifetime I've spent roaming outdoors in places without roads, I've never deliberately followed the trace of something other than my species. We walk slowly and carefully enough that the more curious penguins have time to come over and investigate. One molting Adélie, its crest a tuft of downy gray feathers, toddles right up to me, throws its flippers into the air, and starts squawking. The flailing bird's feet look out of place on its body. They are pink, like the inside of my arm, and scaly—prehistoric, even—with each toe ending in a long black claw that grips the rose-stained snow.

Something Gui taught me: the more krill penguins eat, the pinker their poo. The sides of my tongue curl from the tart taste of iodine, released by the dried effluvia lofting into the air. A long line of penguins has formed along the highest ridge of this hill. One by one they waddle over to the edge and peer into the roiling surf, but none takes the plunge.

Soon the slope steepens, and I jab the end of my ax into the snow for traction. The various traces collapse into a single clearly defined trail as we draw closer to the island's highest point. There is, it seems, just one way to move across the slickest section of the ice field. Our safety in this moment depends upon several things: the collective intelligence of the penguins that went this way before, the pickaxes we carry in our hands, the Gore-Tex soles on our feet. Almost every aspect of our mission is threaded through with petrochemicals: not just our means of transport, but these shoes. My camera and tripod and carrying case. The voice recorder where I store my shipmates' stories—many of which, I will learn, start with a parent whose job it was to pull oil from the earth. Nearly all of it, right down to the families that raised us up, runs on petroleum.

Even the child Felipe and I wish to welcome: this too is made possible by unearthing the decomposed bodies of plants and setting them aflame. Without the energy they provide, the energy we only recently started to transform into real speed, my husband—who loves the place he is from wholly—would probably have never left Colombia to study Latin American literature in the United States. Not if it meant that going home would have been difficult. And then, of course, he and I would have never met.

The higher we trudge, the stronger the wind becomes, until eventually we emerge onto a rocky outcropping. The world that lies before us dense with light. No trees or bushes or scrub. Nothing green anywhere. Just white of ice, blue of water, and gray of granite. Over fifty-six million years ago, forests flourished at these southern latitudes. At that time, the earth was, on average, seven or eight degrees warmer than it is today. CO_2 levels were one thousand parts per million, a concentration that—if we continue with business as usual—the IPCC suggests we might reach sometime near the end of this century. When the children I hope to have will have grandchildren of their own.

"Now those are some well-defined beaches," Scott says. They are not the white-sand, piña colada–sipping kind of beach, but a long string of rocky terraces. "I need a couple of minutes to count them and come up with a plan of action."

James pulls out his massive telephoto lens.

"Can you take a couple for me?" Meghan asks. "Of the Adélies? If all I send home are photos of rocks, my family's gonna be pissed."

Kelly apologizes for snapping at me down at the landing site. "It's just that I'm tired of being under constant surveillance," she says. "You're not like the rest of them, anyway." She gestures to the *Palmer*, where the other two journalists remain, logging their stories.

"I'm trying," I say, extending my palms.

"I know that."

"You all are taking the time to be interviewed. I figure I ought to return the favor." The silence that follows feels more companionable, and I go back to marveling at how the sea, the sky, and the stone melt together to make a complete picture, a symphony of sorts, that was, for the longest time, beyond our influence.

"Let's divide into two teams. James and I will start in the middle and work our way up. You guys start on the beach below ours and work your way down," Scott says. "We literally only have one data point from this area, so anything you can get your hands on will help."

We step out of the windbreak and begin down the stony incline. The cold amplified by the bursts of air barreling up the slope. As we go, I ask a question that has been on my mind ever since I learned how much of our cruise is committed to reconstructing the past. "What happened here over the past couple thousand years isn't what will likely happen tomorrow or in a hundred years' time. That's a big part of what climate change has altered," I say. "So why are we trying so hard to learn about what took place?"

"Once you know something about what things used to look like, you can start to add information about what we're seeing today," Kelly says. "Otherwise, you can kind of make a model do whatever you want."

"This is our beach?" Meghan half asks Scott, pointing to a faint ridge of pebbles on the hillside. "And above it is yours?"

"That looks right to me," says Scott.

Kelly cinches her hood, obscuring every facial feature except her eyes, and rephrases her answer. "There's no exact analog for our present conditions in the past. There's never been a moment like this one. But what we can do is try to get at the processes that go on in a 'natural' environment, a naturally forced environment, and see how quickly they move on their own. Then we can look at what is happening now and do a comparison, which will give us a sense of just how fundamentally the changes we're witnessing stray from the previous baseline."

"Let's talk about this once we actually get the samples we need," Scott says, handing me a shovel. "Holler if you need a break."

CINDY:

I was born in Bakersfield, California, halfway down the state. When my mom went into labor, my dad was about a five-hour drive away. He was probably fitting pipe. There were lots of pipelines going in at the time. The oil industry provided a lot of jobs back then. Still does. I mean, that's what Edison Chouest, the company that owns this boat, that's how they make most of their money. They have their big shipyard in Louisiana, where they service the offshore oil industry. My husband, for years he flew a helicopter out of Port Fourchon to the Chevron and Shell oil rigs in the Gulf.

BECKY:

My mom had all four of us naturally and in under three hours each. I was the fastest. I was number four. What I know about my birth is that my birth took an hour and a half. My dad was sitting on wells in western Kansas. He was a geologist for an oil company back then. When he got the call, he drove like mad across the state to get to the hospital in time.

VICTORIA:

Becky's dad was a geologist working in the oil fields, my dad was a swamper. It's somebody who—I'm not quite sure, but they do something with the rig. They set it up and break it down, that kind of thing. So he's out there in the oil fields—they'd go out for two, three weeks at a time—and my mom, she's super pregnant.

She's a waitress, like a diner waitress, at a place called the Kettle. She has the day off when she goes into labor. My grandma or one of my mom's sisters—I can't remember which—takes her to the hospital. That hospital doesn't exist anymore; it's all boarded up. My mom is calling my dad, trying to get hold of him. But how do you get hold of someone pre–cell phone? You have to call out to the company, and they have to call out to someone who gets hold of somebody else. My dad finds out somehow, but he's also way out there. She's in labor, and I guess things aren't going great, because I was born by C-section. She did the laboring by herself because my dad didn't make it back in time.

THE SUN GLITTERS HARD OFF the patches of ice and snow that dot the tiered hillside. On the ship, I'd learned that when a glacier melts the earth beneath it rebounds, like a memory foam mattress. I hadn't quite gotten it at the time, but now, standing amid the terraces of Schaefer Island, I do. As the Canisteo glacier thinned and retreated, this land, which had been beneath it, rose: solid stone pushing up through the ocean and air; the rock itself incrementally responding to the glacier's decreasing weight.

"We're going to dig a series of pits about a foot or two deep and sift through the debris looking for anything organic," Meghan says with a confidence I do not expect. She is quiet on the ship, and young, but out here in the wind, she sloughs all that. "It can be a sliver of a shell, a penguin bone, seal fur, whatever. When we find something, we take a photograph of the sample in situ, bag it, label it, take a GPS reading, and record all that information in the field notebook. The goal is to gather at least three samples per beach." The organic material we collect will teach us the age of each terrace, information which will later be plotted against its elevation, creating a curve of the rebound rate over time.

"Where should we start?" I ask. I pull my head deeper into my jackets before tightening both of my hoods.

"Here's just fine," Meghan says, kicking at a big rock to see if it's still frozen to the earth. It wiggles but only a little. "If we'd arrived three weeks ago, we wouldn't be able to work."

I push the shovel's nose into the gray gravel once, twice, three times, gently distributing the stones in a ring around the growing pit. Kelly and Meghan lie stomachs down on the hard ground, running their fingers through the pebbles I've unearthed.

While technically tasked with the retrieval and analysis of seafloor bathymetry, Kelly helps whomever needs it. "What do you think of this?" she asks, pointing toward a slender ochre thing the size of a peanut.

Meghan shakes her head. "Rock," she says.

I continue digging, and soon everything else falls away—the cold, the ice, even the penguins. My focus becomes narrow and sharp as a razor's edge. The work we do today will, ideally, improve sea level–rise models the world over. Never before has ten hours of my life counted for so much.

The deeper I dig, the harder the ground grows, until it becomes impossible to proceed. Then I move three meters to the left and start a new hole. When the women have finished working through the contents of the previous pile, they walk over, lie back down, and begin their search again. As I labor, I notice, here and there, pebble mounds marking the surface of the land. At first I think they must be the work of burrowing animals, but then I remember that Antarctica has no land-based amphibian, reptilian, or mammalian population. No, these are boulders the glacier delivered and that time and ice undid. Cleft over and over again, until what was once a large hunk of granite became a pile of perfectly pea-sized stones. In another who-knows-how-many thousands of years, this rock will be sand.

"How 'bout this?" Kelly asks, pointing toward what looks like a miniature stick of dried cinnamon about a foot down.

"That's bone!" says Meghan. Both move forward to shelter the shard, cupping their gloved hands around the edge of the pit in an attempt to break the wind. I run over to the old hole, pick up the burlap sack full of Sharpies and Ziploc bags, and scurry back.

"You're the writer," Meghan tells me. "Write the sample number on the bag. I'll read it to you now: SH-19-1B-1."

I relish the sweep of the marker against the shiny plastic, get a little giddy over the magic it enacts. Through this seemingly arbitrary combination of letters and numbers, this fragment of a long-dead bird has become a piece of evidence, an entry in the Antarctic archive, which will help us to better understand the exceptional nature of the present moment. If part of me is mortified by this robbing of graves, the rest flushes with excitement that this seemingly simple act—beachcombing for old calcium—might bear such remarkable fruit.

"Now hold it by the bone for a photo," Meghan adds. It takes me a minute to locate the shard again, that's how small it is.

Kelly removes her leather work glove, pulls out some tweezers, picks the shard up, and drops it in the bag. Then she holds the bag up to the light, triumphant, before tossing it into the sack.

"One down, thirty or so to go," Meghan says.

"It's wild," I say, "that we just know that no one has ever moved any of this stuff." Back when this was an active beach—a hundred or even a thousand years ago—wave action folded the penguin's carcass into the island's cold bosom, and there it has remained ever since. Complete physical isolation from our meddling makes this bone both reliable and readable, the trace amount of radiocarbon it contains an imprint of just what the glacier was up to way back when.

My fingers and toes prick with the kind of cold that proceeds numbness, so Kelly suggests we jog to the crest of the hill and back again. I attempt to trot behind her but slow my pace after almost tripping, lifting my legs comically high to clear the crags beneath my boots. We look like astronauts tromping across space rock, our bulky outerwear turning our motions mechanical. By the time we return, Meghan has dug another pit, so I join Kelly on the half-frozen ground, nose just inches above the earth.

"This is definitely type-two fun," she says, sieving the frost-bitten stones through her gloved fingers.

"What's that?" I ask.

"Oh, you know—something that's miserable when it's happening and awesome in retrospect."

The hours tick by in an endless blur of tweezers out, camera out, slip a shard into a bag and tuck it into the burlap sack. We do not cross crevasses or drill through miles of ice. Instead our hands move with the exactitude of a parent extracting a splinter, pulling the tips of ancient penguin ribs and fish vertebra from the ice-hardened earth. There is something so mundane about it. What we do once, we do again and again.

After two more pits, Meghan suggests a break, and the three of us wander over to the nearest escarpment. We press our bodies against the granite to get out of the wind. Soon Scott and James trundle down.

"We're hitting the jackpot. How 'bout you?" Scott asks. Tiny icicles dangle from his beard.

"Nearly every pit we dig, we discover something, bones mostly," Kelly says.

"It's like a penguin graveyard down there," Meghan adds. Then she lies back into a big fissure in the rock face and folds her arms over her chest like a mummy.

"I'm getting to utilize all those sea-glass hunting skills I developed as a kid," I say.

Kelly pulls out a bag of banana pancakes. I exchange gummy bears and a Twix for a flapjack. "Thank you, m'dear," she says before gnawing at the end of the frozen chocolate bar.

For a while, no one speaks. Each of us consuming the calories necessary to keep going.

"Right," Scott says, and the five of us stand.

Back on the boat, each bit of organic material we gather will be cleaned, baked, and separated into two identical vials, each carrying the same sample number. The vials will pass through customs in Chile. There, the paths they take to the United States will diverge: one will travel by sea, the other by air. Should something go wrong in transit—plane crash, shipwreck, or simply misplaced cargo—the redundancy ensures that some part of each sample will eventually land at the Woods Hole Oceanographic Institution for radiocarbon dating.

"Think about it," Meghan says later, in the lab, while folding tinfoil boats on which to roast the samples. "Each tiny shard probably cost tens of thousands of dollars to collect."

As the day wears on, we descend into a series of storm beaches, each made hummocky likely by some massive disturbance in the more recent past: an iceberg crashing against the shore or a small tidal wave in the wake of a glacier's calving. The ridges heap up messily, tight on top of each other, making it difficult to distinguish individual beaches from the imprint of singular, destructive events. It does not seem to be a coincidence that the newest beaches are the ones that bear such scars, that the closer we draw to the present moment, the more violent the ice's movement.

Deep in the bottom of one hole, I spot a rind of seashell no larger than a toenail clipping. Prized for their accuracy, these mollusk remnants record actual—as opposed to apparent—ages for the beaches. Kelly tries to scrape the frozen dirt away with a tweezer tip. Then I pour out a little of my warm mint tea, and the ground around the red shard loosens. She gently wiggles the thin scrap until the earth releases its grip. I hold out the sample bag, but just as she lets go, the wind gusts.

"Shit. Shit. Shit," I say as the shell blows away. It is the first we have discovered all day, and now, due to my error, it is gone.

"Happens all the time," Kelly says, immediately returning her gaze to the bottom of the hole. "Look, there's another." I unscrew my thermos lid and begin the process of extraction all over again.

On the third-to-last beach, Meghan holds up a weathered green piece of canvas the size of an envelope. "Someone was here before us," she says.

I laugh, happy to let the idea that we are the first people to set foot on this island dissolve into the ungodly cold air.

Afternoon gives way to evening and eventually to night, the sunlight cycling from bright incandescence into a shade more crystalline and warm. We return to the escarpment and eat disappointingly thin tuna sandwiches. I give away two of my pigs in a blanket, and scarf down the third along with a half a tub of Pringles and another chocolate bar. When I mention my feet have gone numb, James digs around in his backpack and hands me a soft pair of knee-high socks. Collectively we daydream about what Ariel might be cooking up for midnight rations.

"Mac and cheese?"

"Fried rice," says Meghan, inhaling deeply.

"Remember the salmon Wellington?" I ask, my saliva ducts swelling.

Once all fifteen beaches have been sampled, Scott calls over to the *Palmer* on his radio and asks to be picked up.

"Seal . . . first," says Luke on the other end of the line, half his words eaten by the wind.

"We got what we came for," says Scott. "Take twenty minutes to, you know, actually *be* in Antarctica. But don't go too far." Then he promptly saunters a hundred yards across the scree and sits down. James heads to a tumble of rocks beyond our eating ledge that looks like a set of crumbling dentures. This time, when I hit the apex of the island, I keep going, drawn into a valley on its far side where skuas wheel above a frozen lake. Here the land appears serrated, the stone stacked in long ragged rows. Clumps of molted penguin feathers, like large dust bunnies, tumble across the tundra.

The Canisteo glacier, which lies just beyond the thin filament of indigo water circling the island's backside, beckons. The closer I get, the cooler the air becomes, a warning that I am entering its domain. The wings of the two birds circling above are bronzy and big. Soon they start swooping lower and lower, preparing, perhaps, to make off with my red knit cap. After just a few more minutes of walking, I look back. Where I expect to see James perched, snapping photos, there is but a single penguin, the fur on its belly pink with excrement. I am alone for the first time in nearly a month, but I don't enjoy the sensation. Instead a self-interested spike of fear lodges itself deep in my stomach. My throat goes dry.

In Antarctica, to be unaccompanied is to be vulnerable; better to be one of two. My endless trainings have taught me as much. I scramble back toward what looks like the island's forlorn peak. If the *Palmer* set sail, I would freeze, if I didn't starve first. The skuas, they'd eat what was left of me. My bones mingling with the desiccated Adélie remains cracking beneath my boots. Half jogging, breath coming in cold, shallow bursts, I work my way up. Behind me, the glacier withdraws in the opposite direction.

Kelly's outline, backlit against the endless sun, comes swimming into view. On the far side of the hill, I hear Scott and Meghan laughing. As quickly as my fear arose, it subsides.

My footfall becomes more regular. Eventually stops.

I turn toward the sun that will not set and watch as cool, jeweled waves crash against the icebergs stranded in the bay. The light thrown from the nearest star gilding the rugged edges of the giant glacial dome from which these bergs were born. I get the sense that all afternoon I have been eavesdropping on a conversation that has been taking place over hundreds and thousands of years, a conversation whose language is material, written in ice and rock and bone. And while I lack the ability to decipher it, at least I helped to create a record of the exchange, one that a handful of glaciologists will decode, to the extent they can, when we make it back home.

I pull out my notebook and write a few lines about the whistling penguins and the tabulars in the bay. It is the last page, and so I force my script as small as I can make it without removing my glove. The last line reads: *The sun hovers over the frayed edge of the islands, the snow cliffs* . . . and then space runs out, and I let go of trying to render the experience in a language that denies agency to the birds and the sun and the land by refusing them the pronouns that denote personhood. In just a few minutes I will leave all this behind, while the ice will go on changing. Here, and then not, exactly. It will become smaller and smaller, merging berg by berg with the surrounding ocean, which will carry it north, across the Antarctic convergence, across the equator, until some part of all this arrives in the northern hemisphere and puts a little more of the place I call home underwater. It will draw nearer to me, just as I have drawn nearer to it.

Part Three | *Between the Past and the Future*

SETTING: After getting back to the ship and peeling off all those layers, after taking them to the Baltic Room and removing the penguin poo with a pressure washer, after eleven hours of sleep. After Jack makes a heart-shaped king cake for Valentine's Day, which he unveils in the mess hall at midnight. After cutting into those braids of sweet dough and cinnamon covered with lewd sprinkles, after Becky puts on "Mardi Gras Mambo" to dance. After a photo of Bastien with a Weddell seal is tacked to the whiteboard and captioned "sealed with a kiss." After all that, the ship finally draws close to Thwaites.

PETER:

Lars and I just had a look at the map, and we think that Thwaites is probably about twelve hours away. But given how things are going on the cruise, that means that we won't arrive for at least eighteen, if not twenty-four, hours from now.

ALI:

Compared to the Canisteo glacier, Thwaites is going to be a monster.

KELLY:

It won't look like the dome of ice we could see from the islands, which seemed steep even though it probably wasn't. Instead Thwaites will be one big wall of ice, like in *Game of Thrones*.

PETER:

I can't wait to get there, but the thing is, I've learned that the moment you put a time on anything, you fuck it to oblivion. We've been on the ship for—I don't know—three bloody weeks, and we haven't got there yet.

RACHEL:

My shift ends at noon, and then I usually go to bed an hour later. As much as I want to stay awake for our arrival, I just can't. But as soon as I'm up, I'm going to run to the top deck, take a picture, and then politely ask Thwaites to tell me all its secrets, so I can write my PhD.

PETER:

I've been dreaming of the whole champagne-on-the-top-deck fantasy, with string quartets. I have these visions of elegantly sliding along Thwaites with a glass of bubbly, you know, and violins.

RACHEL:

We just passed 74° south.

ALI:

Becky has a twinkle in her eye. That's all I can say.

BECKY:

We're stoked. We're gonna get a good survey. Then we're gonna get some CTDs. And the oceanographers are gonna give me a filter for diatoms. We're going into a blank space—not just any blank space, a really important one—and we're filling it in.

ALI:

I think *stoked* is the word. It seems like it's taken a really long time to get here.

BECKY:

Because it has.

ALI:

Perception and reality align.

Shortly after returning, I learn that in 1947—the same year the first photos of Thwaites were snapped—Jennie Darlington and Jackie Ronne became the first two women to overwinter on the ice. Jennie was twenty-two and recently married to Harry Darlington III, a Princeton dropout who had enlisted as the pilot for one of the last private missions to Antarctica. Together they had sailed to Texas for their honeymoon, and Jennie enjoyed the adventure so much that she continued on to Panama and Valparaíso, Chile, where Harry rendezvoused with the remaining members of the Ronne Expedition. But the expedition's preparations were soon consumed in crises: half of the sled dogs died, their stores were ransacked, and worst of all, the commander, Finn Ronne, wanted to bring Jackie, his wife, along.

On the night before the men were to depart, seven members of the crew threatened to quit because they feared "it would jeopardize [their] physical condition and mental balance" to have to live alongside a woman on the ice. Surprisingly, the mutinous group proposed a compromise: that two ladies are better than one. In her memoir, *My Antarctic Honeymoon*, Jennie recalls Harry telling her, "There are some things women don't do. They don't become Pope or President or go down to the Antarctic!" To her husband's outburst, Jennie "intuitively" said nothing at all. Her silence a sign that her going, or not going, to Antarctica was a decision Harry alone could make. But, she writes, "Inside me something hardened. . . . His words had sparked a determination to prove myself."

Weeks later, the twelve-hundred-ton, diesel-powered *Port of Beaumont* arrived at the United States' East Base on Stonington Island (named after Nathaniel Brown Palmer's hometown). Jennie, a Chanel N°5 and Bergdorf Goodman devotee, describes herself "manhauling" sledges full of rations from the ship to the base, as well as clubbing seals and eating their livers for dinner. She and Harry slept in a little room at the end of the bunkhouse, the door "more for slamming purposes than privacy." A member of a British expedition stationed on the same fifty-acre island later mistook her for a mirage because he hadn't seen a woman in over two years.

Then Jennie became pregnant.

To the best of my knowledge, this is yet another Antarctic first, but one that often escapes the history books. When Jennie finally worked up the courage to tell Harry, she writes, they finished off their remaining stash of cornflakes and Scotch and talked dreamily not of baby names or nursery decorations but of green salads, artichokes with hollandaise, and champagne. For a few days, Harry lied to his colleagues and claimed that Jennie was sick so he had an excuse to bring her breakfast in bed. Other than that, life at East Base continued as normal. Jennie hid her growing belly beneath her many layers of woolen long underwear and continued to take afternoon treks up a glacier to her favorite perch above the frozen bay. Privately, though, she gave her growing babe a sporty nickname: Penguino.

A little over a year after Harry argued that women don't belong in Antarctica, Jennie gazed out over Neny Fjord and realized that things had shifted: "The Antarctic had equalized our relationship. Even the ice was my friend." Before the expedition, she was jealous of Antarctica because it held incredible sway over Harry's heart, but ultimately it not only brought the two of them together, it encouraged them toward a horizontal rapport uncommon in heterosexual couples at the time. Still, when the pack ice showed no sign of opening deep into February—its persistence potentially demanding the expedition spend another winter in Antarctica—the station doctor sent a covert message to two Navy-run icebreakers working in the region.

Since he had not expected a woman to join the expedition, he had no medical supplies to attend to Jennie's impending labor. "Before the icebreakers leave," he begged, "they must be called upon to break the *Beaumont* out." And so the first Antarctic pregnancy, a pregnancy without complication or worry, entered into Antarctic history as reasonable cause for medical evacuation. Of course, calling it a crisis came with the side benefit of freeing the doctor and crew as well. About a week later, the lights of the M/S *Burton Island* and *Edisto* appeared at the northern end of Marguerite Bay, cracking the ice so that Jennie and the rest might return home.

ALI:

I was actually chatting with some of the Swedish team
about AUV plans. Somebody brought me over the
logbook, where Rob had written, *Heading north*, and I

was like, *Huh, hang on, strange.* I looked on the monitors and confirmed that we had indeed turned around. Rob came in, and I could tell from the look on his face straight away. I said, *Is it bad news?* and he just nodded.

GUI:

We were in the Dry Lab, Lars and I, working. Then Rob came to tell us about the change in plan. I thought, *I hope whoever's in trouble isn't too bad.* That's the first thing that came to my mind.

ALI:

Then we were all running around the ship, looking for all our teammates to make sure everyone was accounted for.

LUKE:

I got a phone call to turn the ship around, so I turned the ship around and headed north, to Rothera. Plotted it and ran north.

PETER:

I'd gone to the gym. I'd gotten all refreshed, taken a shower. I'd been sitting for ten hours without much to do, so I'd snuck off to work out, and when I came back, I said, *All right, guys! What's next?* Wrong mood. Then they told me.

"The ship turned around fifteen minutes ago, set a course for Rothera," Rob says to Jeff Goodell, the journalist covering the cruise for *Rolling Stone.* The two men are leaning against the map table in the center of the Dry Lab. I stop transcribing, push my chair back, and walk over. Though I don't like to butt into conversations on the boat, anxiety compels me toward them. "They have a doctor there who's probably young and fairly inexperienced, but the plane comes from Punta twice a week," Rob continues. He sounds deflated.

"What's happened?" Jeff asks. "Is everyone all right?"

"There's been a medical emergency, and we need to evacuate the patient." It is strange to hear him refer to one of our shipmates as "the patient." Before either of us can ask another question, Rob adds, "I really don't know anything else."

There are three EMTs onboard the *Palmer.* For their training not to be enough means the situation is serious. The words of warning my program officer issued when I signed up for the mission crawl through my mind like a ticker tape headline: *It's easier to send help to the space station than it is for us to get help to you.* A burst appendix? Brain aneurysm? What are the ailments for which a few weeks would be too much time? Perhaps there was an accident. I think about Mark, who spent the entire night hauling equipment to redeploy a mooring. And Joee, who just yesterday, was in the Zodiac when the lifting gear failed. "I could have easily suffered a life-changing head injury," she said when we spoke. Maybe the ovens backfired as Scott was drying samples. Or maybe something went wrong in the engine room. What could possibly demand that we turn our backs to Thwaites?

I start to look around to track who is here and who is not. Lars and Rachel and Kelly. Check, check, check. Aleksandra and Tasha. Check, check. James hovers by the electronic map, tracing the distance between us and Rothera Base on the Antarctic Peninsula. "One thousand miles," he says. "Give or take." Peter, check. Bastien, check. Bastien announces that all the scientists are safe. And all the media folks are here. That leaves the marine technicians, the electronics technicians, and the ship's crew.

In my porthole: a ghostly gray berg drags at the horizon. What just moments ago seemed like a sentinel sent out by Thwaites to greet us turns into an obstacle standing between *here* and *safety.*

Gui stumbles toward the map table. I examine his expression to see if he knows anything I do not. "I hope we can help them to be safe as soon as possible," he says, dropping his voice to a whisper.

"Me too," I say. "Me too."

At the midday meeting, Rob stands in the middle of the Dry Lab and makes a long series of announcements.

"We're in the middle of a medical evacuation. The ship turned around at 10:53 a.m., setting a course for Rothera Base. Rothera has an airstrip. The plane will take off from Punta Arenas, land at the base, and return to South America with the patient. The question right now is: Do we have enough fuel onboard the *Palmer* to use all four engines instead of just two? That would increase our average open-water traveling speed from ten knots an hour to thirteen. The engineers are doing the calculations, and we've sent the query over to ASC [Antarctic Support Contract] because we need their approval. Fuel is costly, and we might not have enough to get back if we burn extra going to Rothera."

No one says a word.

Rob continues, "Will the cruise be extended to make up for this lost time? That might happen. I don't suppose they'll grant us nine additional days because the sea ice will close in and we won't be able to work, but do start thinking about whatever arrangements you would need to make if we were to return three or four days behind schedule. Finally, don't expect to get off the ship at Rothera Base. The wharf's being rebuilt, so the medevac will also include a Zodiac ride."

Cindy walks in and unzips her blue down vest. She flips her radio to silent mode. On the mainland she runs a farm, a role she told me requires helping big animals to navigate big changes, which might be why she makes me feel calm. Or perhaps it's because she is a grandmother of five. When I see her soft expression, I remember a sign tacked to the bookshelf in her office: *Think like a proton, always positive.*

"First off, everyone needs to know that, right now, the patient is stable," she says. "We do have enough fuel to run on all four engines, and ASC has approved the decision. So we should get to Rothera in a little under four days."

That's when I know: Lindsey. Lindsey is in trouble. Were Lindsey well, she'd be making this announcement. She'd be working with the crew to do the fuel calculations and contacting ASC. My mind runs in reverse. The last time I saw her was at dinner the previous night, I think, and she had seemed well enough then. Perhaps she fell down the stairs in a swell and broke a hip. Maybe her clothing got caught in a winch. The thought makes my mind contract.

"Is there anything we can do?" I ask.

"Right now, we need you to respect the patient's privacy. Until we've had the chance to contact their family, I ask that you do not speak to anyone off the ship about what's happening. Once we reach the patient's kin, then Peter West, in the Office of Polar Programs, will make an official announcement. We'll let you know beforehand so you can tell your own families that you're okay. And please make sure to limit those phone calls to less than ten minutes; lots of folks are going to be trying to contact home."

LINDSEY:

I wake up in the middle of the night to go to the bathroom, and when I stand up from the toilet and start walking back to my bunk, I feel like a muscle has ripped or torn between my leg and my body. I'm like, *Oh gosh, that's kind of weird.* And then that pain progresses—it moves into, becomes, a kind of pain that I can't describe but that's extreme.

Carmen and I had not necessarily been trying to get pregnant, but we were also not *not* trying to get pregnant. So, you know, there's always that chance. I wake up in tons of pain, and it does run through my mind to start calculating days. I take one of the pregnancy tests we have onboard and it's positive. At this point, I'm like, *Okay* [*voice drops*], *well, that adds another dynamic to whatever this terrible pain is.*

It's also the middle of the night, and so I'm wondering whether I ought to call someone to come to the room. It's weighing on me very heavily that the moment I share this information is also going to be the moment that I destroy the cruise. It seemed like we were never going to get there, and then it seemed like we were so close to Thwaites. And that, that is also the moment when this pain arrives in my abdomen.

Carmen and I have worked down here a lot, and people often don't know that we're married, because Carmen is very work driven when he's at work. I would like to say the same about myself. I'm very work driven as well, but, you know, I would probably be a little less—I don't know—I would probably be a little more affectionate than Carmen is, in a way. So, no, we don't share a room, but that's mostly because our watches are so different. Our sleep patterns and stuff just don't really line up when we're at sea.

Carmen is on watch, and I'm just hoping that the pain would magically go away. But that isn't happening. After an hour or two, I call him and tell him what's going on. I say, *You know, I think I should call Jennie, because she's the EMT.* But I also know the likely outcome of that decision; the outcome will be that the ship turns around. At this point, Carmen is like, *If any other human being came to you with these symptoms and this much pain, you would not hesitate for a moment to call the EMT and get this process started. Just because it's you doesn't make it a different decision.* And of course he was right about that.

Jennie comes. It's about three o'clock in the morning now. This all started at midnight. We call the Palmer Station doctor. He is—you know—that's your resource, when you have something that's above your head you call one of the station doctors. We know the Palmer Station doctor personally, so we call him and tell him what's going on.

As I describe my symptoms, he is all deep sighs and long pauses. Finally, he says, *If you were at home right now, I would tell you to go immediately to the emergency room.* And I say something like, *Yup, mm-hmm. That's not an option. But we're all on the same page there, that's what would be happening if we were somewhere else.* As an EMT, I had already gotten to the worst-case scenario in my own mind: that an ectopic pregnancy could rupture and . . . kill me [*laughs*] . . . while I'm at sea.

Prior to the cruise, I sat in on some meetings about what would happen should an emergency arise while we're at Thwaites. We ran through the options many times, because the research site is extremely remote. I already knew that we were likely five days from Rothera, and that Rothera is the base we would go to should the need arise.

So I'm feeling the doctor out about what he thinks of Rothera being our best evacuation option, and he says, in not so many words, if all of these things are pointing toward the worst-case scenario, then five or six days is actually [*laughs again*] too many days.

That's like . . . okay . . . okay . . . that means we'll make it or we won't. I don't know.

If the ectopic pregnancy causes my fallopian tube to rupture, there's nothing we can do. I ask him, *How will I know if this is happening? I'm in extreme pain* right now, *am I bleeding out into my pelvis* right now? He tells me that I should push on my pelvis to see if it's getting hard. That my abdomen will get distended if it has that much liquid in it. Of course, I immediately start palpating my pelvis. And am deeply thankful that it's squishy.

There's also a part of me that's like, let's talk *not* worst-case scenario. Let's talk most ridiculous scenario:

that I have extreme gas, for instance, or am horribly constipated. So simultaneously I also start working from a place of trying to fix what could be a small issue. I go ahead and take some laxatives. I'm thinking I have a few hours to make this better. So I did do that. I took some laxatives.

At this point, I'm very scared to go to sleep. For some reason, my brain is like, *What if this happens, and I don't know this is happening?* I mean, what do you do? Nothing different, I guess. I ask—'cause this is the middle of the night, and even more so the middle of the night in Denver [where ASC is headquartered]—I ask him, *What are the proper channels we need to go through to make this a medevac?* Because, at the time, we're still going full speed ahead toward Thwaites Glacier. I tell him I think we should get the ball rolling on turning the boat around, and he's like, *I think we should wait until the office opens in the morning, because the number of hours we have isn't going to make a difference on the outcome.* [*disbelieving laughter*]

Of course, I'm immediately back in my cabin on my eighty megabytes of phone usage, trying to Google *ectopic pregnancy* and thinking, after waiting for forever for the page to load, that this is bad. Ectopic pregnancies are so rare and so fixable. It looks not bad at all if you're in an area where you can go to a doctor, but if you're where we are, it looks bad.

I continue to be in pain powerful enough to call the doctor back and say, *I can't wait until whatever time we discussed. The pain is continuing to hold, and it's really worrisome. I want to do something now.* And so we do. I call Denver and start the evacuation process.

> I'm not very religious, but my mom really is, so for whatever reason, in that moment, I was like, man, my mom has got this. I'm just going to call her and tell her what's happening, and she will broadcast this to every prayer group and church group on the planet to get some good vibrations going my way. I told my mom: *Please just get me home.*

DURING THE SAME STRANGE SEASON in which I read *My Antarctic Honeymoon*, I write to Lindsey. I hadn't interviewed her during the medical evacuation, out of respect for her pain and privacy. But now that half a year has passed, I ask whether she would be willing to share her experience with me. She is back in Punta "babysitting" the *Palmer* and suggests I call the ship. How can I describe what it is like to be on the other end of the phone? I am elated, for lots of reasons. First, that Lindsey is well. Second, that she is pregnant. Third, that she is willing to tell me about it.

"Gosh," she begins. "Thwaites is so far away in my mind because of everything that's happened since then. You know, Carmen and I met working down here. We knew each other for years before we started dating, which also happened on a cruise. Antarctica is our little love story, and now we have a little love baby."

"Its body is made from Thwaites's meltwater," I say, thinking about the desalination system on the ship and how the drinking water it provided us also fattened the cells of the embryo: the outer layer blooming into the brain and nervous system, eyes, tooth enamel, skin, and nails; the inner layer, the digestive system and lungs; and everything in between, nothing less than baby muscle, bone, and heart. Heart that circulated blood between Lindsey's body and its own, heart that pumped milk from a great glacier between mother and soon-to-be-born son.

"We're actually wondering if we should use an Antarctic name for this boy. Or is that too weird?"

"Oh, I love it," I say. "The continent would finally be naming one of us."

"If we call him Thwaites, though, I'm afraid there will be a lot of eye-rolling, and that he'll have to tell the story again and again about Mom and Dad getting evacuated from Antarctica."

"Never, not once, did it occur to me that we were leaving because you were pregnant." As soon as I say those words, my elation ebbs, replaced by a new concern. Perhaps Antarctica and motherhood weren't meant to mix after all, perhaps that preparatory pelvic exam wasn't as misguided as I originally thought. "Felipe and I are trying too," I finally say, changing course. "Funnily enough, yesterday I woke up in pain and thought, *Maybe I'm pregnant*. Then I thought, *Maybe it's ectopic*."

Lindsey doesn't offer platitudes about letting go of control or leaning into the unknown. Instead, she says, "Ectopic pregnancies are easily caught and treated when you can actually get an ultrasound. I swear, it's a never-ending stream of worries the whole time we're brewing these little suckers, which"—she pauses—"I guess will likely last for the rest of our lives?"

I laugh. Grateful for how Lindsey includes me in the new-parents club, even though neither of us knows what's going on in my uterus. For a little while, it feels as though I am back on the ship, the conversation easy and unrushed. We cover lots of ground, from Ali Wong to Ursula Le Guin to how crazy it is that babies come with their own yolk sacs. The next morning, I wake with bloodied underpants again. I am sad but a little less so this time because I know that, even if I don't get pregnant, this story will have at least one new mother in it.

ALI:

The science is why we're here, but we're all human and we all understand: if I was in that situation, I would want the ship to turn around and take me home.

JOEE:

We've all been on the ship together for a little over three weeks now. Has it only been three weeks? It feels like three months. But even in that short time, we've developed an intimacy and a sort of rhythm to our daily lives. People working on a ship together quickly become invested in each other and each other's well-being. So the medevac is a pretty weighty thing for the shared psyche of the ship.

GUI:

To be honest, I'm just trying to be as respectful as I can. You know: I'm not asking any questions. The only bad thing about this attitude is that I won't be able to say goodbye to the person, because I don't even know who it is.

BASTIEN:

Some people don't know Carmen and Lindsey are together. They hide it pretty well, but they have those silicone wedding rings, you know, the ones for people who work with their hands.

JOEE:

I'm not acting as an EMT. I have plenty to focus on in terms of caring emotionally for the patient and for others onboard; I don't need the additional responsibility of being the point person in a potential catastrophe.

PETER:

You can't do anything other than get people safe. In open water, we're doing thirteen knots. It's a good clip for this ship.

LUKE:

We're getting through the ice—that's our main concern—then turning eastward. ETA is hard to say; that's captain's call. I can't give any ETAs.

CINDY:

If we stay calm, if the captain and the mates stay calm, it trickles down.

AT FIRST, A TACITLY AGREED-UPON hush envelops the *Palmer.* It is as if we are collectively holding our breath. Time turns sticky like taffy, as we spend a solid day and a half working back through the very same ice fields we crossed to get to the inner edge of Thwaites. John Carpenter's Antarctic horror film, *The Thing*, loops on the shipboard television. For a little while, folks hibernate, retreating to their rooms to read or sleep. I write an article for *National Geographic*, finish season two of *This Is Us*, read the first three chapters of Adrienne Rich's nonfiction book *Of Woman Born*, and move on to Richard Powers's novel *The Overstory*. Along with Barry, the creative genius behind the communal crossword puzzles, I also set about organizing the International Amundsen Sea Ping-Pong Tournament. First, we post a sign-up sheet in the computer lab, and when we have nearly half of those onboard interested, Barry fires up the old Epson and prints a poster-sized bracket.

"I've never seen a cruise actually complete a tournament," Barry warns as I write each of the participants' names on slips of paper that I plan to pull from my beanie to determine first-round pairings. "Getting people to actually play is like herding cats." We're sitting in a little cubby on the far side of the computer lab: the walls covered in circuit boards, spools of colored wire, and cans of condensed air. Stickers from Antarctica's research stations are tacked to the cabinet, along with a photo of a bunch of techs wearing float coats and nothing else. Barry notices me looking at it. "We did that in protest," he says, "when ACS threatened to stop supplying us with Carhartts."

"They wouldn't."

"They did."

Barry rubs his calloused palm over his thigh and picks up a two-inch-tall purple plastic penguin.

"Can we offer that as a prize to the winner?" I ask.

"It's not mine. Another tech made this on the 3D printer. It belongs"—Barry searches for the right words—"to the ship."

"Could we make more?" I imagine an elaborate awards ceremony with emperor penguin trophies.

"I don't have the code. But I could donate something else, from my personal stash," he says, tugging at the brim of his Tennessee Ridge cap. We hang a canvas sack in the hallway with a small sign. It reads: *Donations for the Winner Welcome.* All sorts of special prizes appear: a bag of licorice, a flip knife with a macramé key chain, chocolates with cherry liqueur in the center, and something marshmallow derivative. In those odd, long days of passage, we mother each other, inventing diversions that double as displays of care. Bastien founds a "Sauna Club," which convenes at four-thirty each afternoon to help us work up an appetite for supper after yet another twenty-four hours of not doing much. Anna takes over evening entertainment.

The first time I show up for one of her bridge lessons in the conference room, there are white plastic bowls full of Swedish candies placed at strategic lengths down the center of the table. Bastien picks up a little black licorice sailboat, pops it into his mouth, chews, then walks over to the trash can and spits out the half-masticated lump.

"The snacks are for shit," he says. "But I'll stay anyway."

Unflappable Anna is not offended; her excitement at the possibility of holding a proper match, contagious. "Each round consists of three stages: the auction, play, and scoring. Bid one if you have at least twelve points and four cards in that suit, and so on," she says. I look at Bastien, who shrugs and looks back at Anna. "We'll play open handed at first, until you guys get the hang of it." The next night Bastien and I team up. We bluff, partly to encourage the others to bid higher than they ought to, partly to entertain ourselves. After days of worrying about whether one of our shipmates will die, it feels delicious to do something both low-stakes and risky. During the auction's last round, the other team figures out our ruse, and to call us on it, they double down. I give Bastien a look that says, *We're sunk*, then blow the first play, losing control to the other team.

The night is foggy, the *Palmer* sailing half-blind through the downy heart of the Bellingshausen Sea. Bastien has been shaving his facial hair into strange configurations ever since we turned around. Today his wide, bushy mustache makes him look like the singer in a 1970s lounge act. He runs his thumb and forefinger along its length. The ship suddenly lists to the starboard side. "Iceberg!" he chuckles. Then he leans in, throws a high card, and takes over the deck. As the plays progress, we work the trump, collecting trick after trick.

It's past midnight when I wander up to the physical bridge, still drunk from the game. Luke points to a spot in the distance. I squint but don't see a thing. Right as I am about to look away, a humpback sends up a spout. Out of the ocean comes its fluke in a long, languid undulation. At the top of the roll, the underside of its tail flips up: a lunar crescent in this otherwise moonless white night. The sight of it is calming somehow, this big animal turning in the sea we share. Then the whale descends into the deep and so do I, back to my bunk, to try to sleep.

MEGHAN:

People at home keep asking if we've hit Thwaites, and I keep having to tell them, *Not yet*.

GUI:

We're keeping ourselves entertained with a lot of ping-pong. And there are plenty of beautiful pictures to take. Of whales especially. Yesterday I saw a humpback breach.

BASTIEN:

Have you ever seen *Groundhog Day*? Every day is the same, over and over again. That's what life is like right now.

KELLY:

Jennie's got *The Thing* playing on repeat. It's a cult classic.

ALI:

It's a good B movie. A horror film.

KELLY:

It's better than that.

JOHAN:

I haven't tried bridge yet. It's same with the sauna and other things. Card games require beer to be fun.

SALAR:

What wouldn't I give for a beer, or several.

SCOTT:

I think I have reached my, uh—I don't know what the right word is, but at this point, I feel like it's been a long cruise, and we haven't even gotten to Thwaites yet.

BASTIEN:

My birthday cake was delicious. It wasn't the traditional carrot cake that you find on land. But it was definitely cake, and there was carrot in it.

SCOTT:

Meghan and I are in the labs cleaning penguin bones and drying them out in the oven. If you don't have anything coming up, it's like: What day is it? What month is it? It doesn't really matter. People eat breakfast at midnight—

MEGHAN:

—and their breakfast is steak.

PETER:

Over the next eight days, I'll write at least half a dozen papers. I'll do the acoustic Doppler current profile section of the cruise report. It will be word-perfect by the time we get back to Thwaites. I'll also probably read *War and Peace*. We'll work our way down the list from there.

DURING THE FIRST FALL AFTER the cruise, I contact Meghan Kallman, the cofounder of Conceivable Future, a woman-led collective that brings attention to the threat climate change poses to reproductive justice. Unlike other organizations that link the long-term health of the environment to decisions made today about child-rearing, Conceivable Future doesn't advocate for population control or choosing to be child-free; if anything, they urge elected officials to stop underwriting the oil and gas industry. The collective's power is not born of the conservative logic of "fighting for the children," nor does it rely on the punk allure of "voluntary human extinction." Instead, their argument is messy and grounded, gaining strength from its plurality, from the loose network of people who, by reckoning with the climate crisis's impact on their reproductive options, have become active rather than despondent, deliberate rather than stuck.

As I scroll through video testimonies on Conceivable Future's website, I encounter strangers talking about how their deepest desires intersect with what they most fear—and what they have chosen to do as a result. The pregnant professor explains that, for her, having a child is a twofold commitment: first, to living her life as meaningfully as possible, despite her cynicism at the current state of things; second, to the world her child will inherit, which has spurred her to participate in the fight for climate justice and workers' rights. The Puerto Rican student of conservation biology vows to raise his future children to be both critical thinkers and caretakers of the earth, while acknowledging that being able to make this choice is a privilege not equally afforded to all. The young man who grew up next to a coal-fired power plant plans to get a vasectomy because of the physical burden he carries in his body and the ecological grief that weighs on his heart. Despite how unique each voice is, there is also an uncanny familiarity to it all, a sense of eavesdropping not only on their thoughts but my own.

In addition to all the staccato starting and stopping of speech, the *um*s and eyeblinks, the searching for language to describe the indescribable, what unites each monologue is the speaker's attempt to regain control in a world where individuals so easily feel powerless. I am moved by the moment when, in the middle of Meghan's video, she shifts her gaze toward the corner of the room and tries to find a way to talk about her niece being born on a day when the temperature reached 105 degrees in northern Vermont—how she was born onto an Earth hotter and less stable than any humans have ever before inhabited. Meghan's gaze lingers on the dust motes, because what do you do with that knowledge? How can we act when the things we depend upon have become undependable?

As fate would have it, Meghan is my neighbor. We meet at the vegan bakery in the strip mall between our homes, where she arrives wearing sunglasses, strappy gladiator sandals, and dangly, leaf-shaped earrings. By way of personal introduction, she tells me that she teaches sociology at UMass Boston, serves as city counselor for Ward 5 in the town of Pawtucket, and is preparing to run for state senate. I tell her of my hope to become a mother and also about my recent trip to Antarctica. When I get to the part about the medevac, she asks a simple question: "Didn't they have an ultrasound on the boat?"

"No," I say. "Probably because pregnant people are prohibited from the ice."

"Right," Meghan says, lifting her eyebrow, her voice deadpan.

Her question stays with me long after the caffeine has been metabolized, the word *ultrasound* sitting uncomfortably in my mind. My Google search for "ultrasound" produces familiar information at first: porpoises and bats use it to locate prey, its frequency higher than humans can hear. But I am more interested in the history of these sound waves—how we harnessed this technology to see in the dark, and to what ends.

The Brits sent these seemingly silent pulses into the North Atlantic's cold oblivion at the end of World War I. Nearly forty years later, during his explosive-detonating tractor trip, Charles Bentley began using similar technology to measure the thickness of the West Antarctic Ice Sheet. That same year, in 1957, Ian Donald, a former medical officer in the Royal Air Force, invented the first obstetric ultrasound. Soon doctors would be able to chart fetal development, sex the child, and identify if the parent was carrying more than one. "There is not so much difference after all between a fetus in utero and a submarine at sea," he said. "It is simply a question of refinement." On the *Palmer*, the Knudsen CHIRP 3260 was tuned to pick up changes in the depth and composition of the seafloor; while the sound waves an ob-gyn sends into a person's body reflect bone and flesh. In both cases, when the wave encounters something solid—like rock or skull—the image appears sharp and clearly defined, while softer materials—like sand or foot pads—yield something looser looking, more granular.

I think back to Jennie Darlington. How it wasn't exactly her pregnancy that pushed the doctor to call for a medical evacuation but his own inability to provide her with reliable care. *Ectopic*, I also discover, means "out of place." As in how women are still, at a deep structural level, treated to this day, on the ice and elsewhere. Had the *Palmer* had a medical ultrasound onboard, we would have known enough about Lindsey's condition to keep us from turning around. That single word, *ultrasound*, like the pulses of noise we released into the ocean, reveals what was previously hidden: that it is possible to say it was not the pregnancy that caused our evacuation, but our inability to imagine a pregnant person alongside the "Doomsday Glacier." That it was this historical exclusion that resulted in the ship running circles in the endless Antarctic light.

LINDSEY:

There was talk of having to crane me off the *Palmer* into the Zodiac. I just wanted to be in an upswing where I didn't feel so terrible. I wanted to be able to get down the Jacob's ladder and into the small boat on my own. Because I'm causing all this trouble—I mean, I feel like I have destroyed the cruise—and so I want, at the least, to be able to do that part myself.

By the time we actually reach Rothera, I'm also feeling very relieved that we've gotten that far. They had delayed the Dash 7. It was all packed up and ready to go, so as soon as we got on it, they took off. There was actually a doctor on the plane—not that that changes anything, but it made me feel better.

The doctor in Punta showed us the baby first, which was very sweet of him. He did a lot of looking around my abdomen and not saying much. We were like, *Uhhhhh?* and then he showed us the baby and it was just a little nugget. He let us listen to its heart. And we were like, *Oh, that's its heartbeat!* He told us, *Yes, the baby is very happy, very healthy.*

And then he was like, *You have these* kiests. Carmen and I were like, *Okay. What could that mean?* You could see it on the ultrasound. He is pointing to it, this huge black thing. Finally, he writes the word down on a piece of paper, and we are like, *CYSTS!* He communicates to us that it is normal—well, not normal, but they're not dangerous.

Once he got across the point that the cysts were not going to harm me, we were, like, instantaneously all in on this baby plan, strangely enough. Before, I thought: *Who cares if something happens to the baby, get me to the hospital!* But

> the second you see it and listen to its heartbeat, you're like: *But wait! It's gonna be okay, right?* He told us that the baby was fine, that the cysts and the baby were two separate things and the baby should not be affected. And we were like, *Oh my gosh, this is everything we'd hoped.* And it was. Some weird condition that was not life-threatening and a happy, healthy baby.
>
> Once that fright of possibly dying on the cruise was removed, once I got to the hospital, my focus also shifts really dramatically. I just begin to feel bad for everyone else. I've worked down here for twelve years. I'm very invested and passionate about helping scientists get to Antarctica and collect their data so they can share that information with the rest of the world. And this Thwaites cruise was the best venue for getting that information out, because of the media presence and because of having a writer onboard like yourself. Suddenly I start thinking that I had taken away people's ability to collect the information they need and get it out to the world, and that's also very upsetting but in a totally different way.

BECAUSE OF BAD WEATHER IN Punta Arenas, the Dash 7 that brought Carmen and Lindsey to South America can't return immediately with the replacement technician. If they don't arrive in the next twenty-four hours we will have to leave without them, limiting what we will be able to accomplish should we make it back to Thwaites. At the midday meeting, Rob makes another announcement. "We're in a holding pattern until conditions improve," he says. Then he looks out the porthole and adds, his voice high and indignant, "But it's gorgeous *here*. I can even see the sun." Good days don't come often in Antarctica, even on the peninsula, and there is a chance that another so perfect for landing the plane is weeks away.

I spend most of the afternoon on the bridge with Nancy Campbell's *The Library of Ice*. But no matter how beautiful the prose, it's frustrating to read about representations of snow and glaciers in literature when I ought to be experiencing them firsthand. Meghan sits beside me, working her way through the second novel in Jeff VanderMeer's *Southern Reach* trilogy, and Bastien sprawls on the padded bench behind us, devouring John Wyndham's science-fiction classic *The Kraken Wakes*.

Meghan leans over the armrest. "You know, when I told my boyfriend what we've been doing every day, he said it sounds like we're in an old people's home."

"If we stay here any longer, they're going to have to lace our food with sleeping pills to knock us out until the news gets better," Bastien responds, before returning his gaze to his paperback. Now that Lindsey is safe, the unspoken ship-wide mandate to keep it together is no longer in effect. On this, the fifth day since the medical evacuation began, the mood shifts from solemnity to a mix of exasperation and slaphappy relief. I sigh and step outside. Gossamer strips of snow pull off Reptile Ridge's dark spine. Behind it rises a chain of white mountain peaks so shamelessly dramatic I wonder when, if ever, anyone touched their tops. Barry's got his electric guitar out and is working his way through a twangy rendition of "Hey Jude." Soon a couple of orcas appear off the port side. Just behind them, on the southern tip of Rothera Point, sit half a dozen drab green buildings. A work crane cantilevers over a backup generator. I watch as someone trudges from what looks like an airplane hangar to an extra-large double-wide, making their way past a bunch of oil drums on a flatbed. This is the closest I've been to human civilization in nearly a month, and yet the encampment doesn't make me feel like I've been missing much.

That evening, we are told that, with the additional delay, the British government has mercifully decided to extend an invitation to visit the base. Someone tacks a sign-up sheet to the door by the Dry Lab. As I write my name down, I fantasize about walking at a brisk pace over solid land. The next morning, Rothera sends one of their sporty small boats across the bay. It sidles up next to the *Palmer*, and five of their scientists disembark. The day is flat and calm, and rumor has it the Dash 7 will arrive in the afternoon. I descend the Jacob's ladder to the red inflatable and grab the hand of the bearded, blue-hat-wearing stranger at its base.

"What's with the paint job?" he asks, nodding to the *Palmer*, his lilting Irish accent fresh in my ears. "Was orange on sale or something?" Before I can answer, he slaps his knee, then turns to help the next person board the boat. I stare at him for longer than I ought to, at his goofy expression and sun-blasted freckles. I can't tell if I find his face handsome or just different from my shipmates'.

On land, we are greeted by Mairi Simms, an atmospheric scientist and our guide. "I've spent five of the past seven years here, three winters," she says, wind rouging her porcelain cheeks. "It's the sense of community that keeps me coming back, especially in the winter, when there are only twenty of us. We're the fire patrol. We're the doctors. We need every person who's here." She leads us from the makeshift pier, across the airstrip, over to the Bonner Lab. "Every pathway you walk on, even the runway—all of it's made from rock blasted from that mountainside," she says, gesturing over her left shoulder. I stumble at first, unaccustomed to walking on anything other than a boat deck.

Mairi pauses just outside the building to let us know that our time on land is limited by the fact that nearly everyone wants to see the station. "You've been divided into groups, each of which will get an hour and a half to take in the local sights and return," she says, then she pushes the massive wooden door open. Inside a sign reads "ENSURE YOUR MIND AND BODY IS CLEAN BEFORE WALKING DOWN THIS HALLWAY." Mairi leads us past a cylindrical hyperbaric chamber where they can recompress divers in case of an emergency. We walk past a dozen indigo tanks that hold all kinds of cool Antarctic specimens. There is a forty-armed starfish, a sea spider named Joseph, anemones, brachiopods, and many different snails. One cistern is labeled "temperature +2° C" and another has a plastic scuba diver perched on the edge.

Melody Clark, a geneticist, welcomes us to the Aquarium Room and begins to explain her research into the effects of climate change on the reproductive capacity of Antarctic bivalves. "Clams reproduce by releasing sperm into the water, and that sperm can travel quite a long way before it encounters an egg. So we're looking at the genetics of individual animals—one meter apart, five meters apart, fifty meters apart, and fifty kilometers apart—to see if there's any gene flow between them." Direct observations of evolutionary change in nature are rare because of the time scale at which these stories tend to unfold: over millennia, as opposed to days or months or years. But the Southern Ocean's extreme cold, the rapidity of the changes taking place there, and the very specific adaptations certain creatures increasingly rely upon to thrive make it the ideal laboratory. "The point is that its warmer at the very top of the peninsula, at Signey Island, than it is further south. If the genes are intermixing between communities, then this is an example of warm-dwelling clams helping the colder-temperature animals adapt to climate change through reproduction."

As we walk back down the hall, I think about the climate change adaptation method with which I am most familiar: relocation. We know that the species that can track their thermal niche by moving uphill or poleward have an edge over those that are stuck in place. Now I begin to wonder whether animals with shorter life cycles, or at least less time separating birth from sexual maturity, might also have an advantage: adaptation through evolution. The faster an animal can cycle through generations, perhaps, the faster changes can take effect in the wider community.

"We've found that the younger clams survive much better in changing conditions, which is surprising but also complicates our idea of how these animals might biologically adapt, because it's the older ones that reproduce in larger numbers," Melody says, as if reading my mind. "These animals live an absolute minimalist lifestyle. Some go three years without eating. And most take six or seven years to start spawning. That length of time, between birth and reproduction, doesn't help."

She opens the door to her office and sneezes. I step away, fearful of carrying germs back to the boat. There's a window on the far wall, a single pane framed in burgundy, through which the long, low lip of the Wormald Ice Piedmont is visible. "The glacier behind the station has stepped back significantly since I first came here," she says. "Where your boat is now—even that used to be beneath the ice."

Soon Mairi ushers us over to Bransfield House, the social hub of the base, which looks like a cross-country ski lodge. On our way, we pass an elephant seal sleeping in the sun on a little land bridge. "They adopted that spot years ago," says Mairi. "And Antarctic treaty regulations restrict interfering with the natives, so they get to stay." In the mudroom, we remove our shoes and coats and wash our hands. Once inside, I plod down the carpeted hallway in my wool socks, examining the portraits on the wall. There are no penguins, nor photos of cleaving ice. Instead, I see every single person who has over-wintered here. Arranged by year, the photos reach all the way back to the sixties, a full decade before formal construction of Rothera Base began.

I scan the faces, searching for a set of smooth cheeks among the beards. Some photos are taken outside, making it difficult to distinguish between fur-lined hoods and facial hair. Others feature men in matching flannels. All eight who wintered over in 1973 wear dress coats and ties. There are happy huskies and cans of beer, snowmen and prop planes. "Come to the canteen," Mairi tells me, and so I dutifully follow her farther down the hall, momentarily putting my search aside. The canteen, which disappointingly only serves alcoholic beverages after midday, also doubles as a souvenir shop. My shipmates move around the long tables, fondling all the stuff for sale—neck gaiters, sweatshirts, stuffed penguins—with disturbing concentration. In my wallet I discover a ten and three ones. Settle on seven postcards and six stamps that I am told will send my missives most anyplace in the world.

"You're doing it all wrong," says Lars from the couch in the corner. "You ought to just sit back and have a moment." Then he turns his attention from the group to me. “Did you find them?” he asks, knowing full well of my obsession with women in Antarctica.

“Not yet.”

“1997,” he says, shaking his head.

“I should be shocked, but I'm not.”

I spend my last half hour on land hastily scribbling notes to friends and family. The card I send to my husband reads:

> Love, light, warmth in my Antarctic night. I am at the bottom of the world, but you are here too in the very way the wind picks up the snow and pulls it in long lines of airborne lace off the mountaintops. You are the rock beneath my feet and here in my heart. Hot, burning bright, bright, bright. Always. I love you.

While I have no problem calling Felipe my husband, I sometimes cringe when he calls me his wife. There are few words heavier with connotations of duty, of submission, of stepping aside. But my discomfort fails to take into account that ours is a relationship where we reverse many of the traditional gender roles—he goes to the market and cooks while I put air in the tires and take the car to the shop; he maintains our emotional equilibrium while I unclog the drains. As I lick the stamp, I think about how, in the context of our relationship, the word *wife* means the person who goes to Antarctica for work, while he, my husband, remains at home.

Then it is time to return to the boat. In the hallway, I pause before the image of Alice Chapman, Jenny Rust, and Lucy Yeomans, the first women to spend a dark season at Rothera. Alice's round face is framed by her wavy brown hair. Lucy and Jenny both sport big sunglasses intended to reduce the glare off glaciers. They look pleased and proud, as does most everyone else in the image. When I finally encounter this photo alongside all the others, my grip on the binary that has defined so much of my searching up until this point loosens. I don't wish to diminish the extra effort these women surely exerted to get here, nor the delight they appear to have taken in it. I want my gaze to be wider than that—I want to see and celebrate the compassion that every person who has spent such an isolated season must have felt for the other people in their extraordinary communities.

My calcified anger dissolves a little, leaving me feeling slightly empty, but in a good way, and ready for more. More time on the boat. More conversations with my shipmates. More birth stories and testaments to the all the support that makes our work possible. More tending to their needs, as they tend to mine. More certain that whatever form care takes, its possibility can be reparative—especially when it is given freely, rather than coerced by centuries of gendered or economic expectation. More animated by the idea that exercising compassion, especially for what isn't ours by way of ownership or biology, "is the unceasing, rigorous work of a lifetime," as Maggie Nelson writes in *On Freedom*, although she cautions, "It isn't something one always gets right." Our execution will likely be imperfect, but that doesn't excuse us from trying.

ACT THREE

Part One | *Arrival*

SETTING: Five days later, during the midday science meeting, Rob makes yet another announcement: seventy miles to go before the *Palmer* is back at the exact location it left when the medevac began. He says fifteen hours to the ice front. Kelly hands him a handmade birthday card with four portholes cut into the cover, each revealing a different photo from the expedition. There's Rob in his orange jumpsuit with a practice Kasten core; Jenny Island's gray gabbro dikes in Rothera Sound; the helo deck, with the *Palmer*'s name painted on the landing pad; and, of course, an iceberg. Everyone sings. No one cares that it sounds horrible. There are three different kinds of pound cake in the galley. Together, the author and her shipmates hope.

Out on the bow, the air is dense and almost warm. We have punched through the sea ice to the Amundsen's foggy interior again. I want to honor the little distance remaining between us and Thwaites, and yet I don't really know what to do except stand here. Just off the port side: a half-flipped berg in the shape of a pyramid. It looks like a ruin, something time has partially undone. What rested below the water line waxed away by the heat of the sea. There the berg's underbelly was gradually winnowed until its equilibrium shifted and a portion of what was submerged pitched into the air. The once-sunk ice smooth as glass.

Kiel walks by on his watch. He asks if I am all right.

"We're so close now," I say. He flattens the folds of his blue jumpsuit, nods, and leaves.

Not a minute later, a long snake of icy air makes its way to where I'm perched. It blows up my spine and enters the centers of my vertebrae, ascending through the rungs of my body in one quick swoop. Instead of filling me with cold, the wind coaxes me upward, a kind of eerie invitation. I imagine lifting my right foot, then my left, to the top of the bulwarks. See my arms outstretched. See my legs in the air. Then underwater. One minute living, very much living, the next—well, in the next moment, my mind returns to me, my body my own again.

Snow.

Light flecks of pearl swirl and drift, hovering against the sea's rich velvet, each tumble full of sudden twists and strange moments of suspension. I try to track a single flake but quickly lose it among the many. The ocean eats each one up. I could go inside, put steel between me and the glacial wind, the strange temptation to lose myself in it. Instead I say to Thwaites, "We're in your territory now," hoping that the words ring like the right kind of surrender.

THAT NIGHT, SOUND SLEEP ELUDES me. I wake often, each time hopeful that we have arrived. Finally, around five o'clock, I rise. Shuffle up the four flights of stairs, undog the door by the Ice Tower, and walk out on the bridge wings.

Thwaites's gray margin wobbles in the gloaming.

We wind alongside, entering small coves and rounding odd promontories. Our pace slow, to hold this precarious line. The ice face soft as dunes. The night's new hint of darkness gives way to the bruised light of dawn, and many others appear to watch what each of us has been working toward—for weeks, for years, and, in some cases, for decades—come into sharp focus. We don't talk. When someone wants to say something, they whisper as though we're in a giant, roofless cathedral. We, who have been at sea for so long, finally gaze upon the glacier that has already given us one another. Rick stands attentive at the ship's helm, the captain next to him, steering us along the edges of Thwaites's unfathomable fracturing, its hemorrhaging heart of milk.

LARS:

There is something I like very much about traveling on a ship. It's not like getting on a plane, and suddenly you're there. No, on a ship, you arrive.

BECKY:

The ice margin is so weird and undulatory. It's not at all what I imagined. It doesn't look like the thick, tabular sheet of ice that you typically see meeting the ocean.

ALI:

I've been working on this region for one-third of my life, and I was worried that Thwaites would be a letdown after seeing so many other parts of the West Antarctic Ice Sheet. But this—this is different, and not in a good way. Some bits look totally mashed up.

BASTIEN:

The face is mangled and gnarly.

JOEE:

It looks like the ice was moving in really abrupt and violent ways, like something tumbling got frozen in space.

ROB:

Thwaites Glacier is a critical boundary in the world today. We've been telling everybody who will listen that this is where rapid change is really happening. And now, here we are, actually standing and looking at the thing that's rapidly changing.

LARS:

Look at that pinnacle of ice. A pillar cracked almost all the way down to the water line. Looks ripe to fall off.

ROB:

Looks like snow sliding off a roof.

JOHAN:

Looks like a ski slope.

ALEKSANDRA:

Look at all that blue ice.

BASTIEN:

This is the coolest shit I've ever seen. I think I'll spend the entire day staring at all these cracks and crevasses and strange, surreal shapes.

ALEKSANDRA:

I can't find proper words to describe what I'm looking at. It's amazing, it's beautiful, it's stunning. I'm here since 6 a.m. I can't stop looking.

GUI:

I have a hard time looking away.

MEGHAN:

I have to go take a picture.

TASHA:

I should get a photo of you looking out.

ANNA:

We're in the right spot, really in the right spot.

PETER:

There is this sense like, *Okay, we can start now.*

BECKY:

It will bend and bend, and then it will break.

MEGHAN:

I've—well, none of us have ever seen anything like this before.

RACHEL:

The captain asked me if I feel like an explorer because we're the first people ever to come here. I was like, *I guess.*

I don't know. That's a very bold statement to make. Someone made my coffee for me this morning. We have a sauna onboard. I *guess* I feel like an explorer. But it's different than it used to be, and that's okay.

GUI:

I just stood outside all morning, staring at everything. All the time I was thinking, *What is all this?* Sometimes I was thinking, *How long is it going to be like this?* And, always, *What is my part in pushing this system to this point, like, what the fuck are we doing?*

NEARLY EVERYONE ONBOARD SPENDS THAT first day up on the bridge in the shadow of Thwaites. We stand together in the difficulty of it, trying to see what sits right in front of us. A slab cantilevers out over the water like the scalloped shell of a giant clam, studded with icicles formed during the recent warm days. I set up my camera to take a series of time-lapse photos. The shutter opens and closes, opens and closes. Art critic John Berger's famous phrase "Seeing comes before words" rises to the surface of my thoughts. The gaze, he argues, is a reciprocal thing; to see is also to be seen, to imagine yourself in the eyes of the other. How do we appear from the glacier's perspective? I struggle to fathom it. The morning we cruise past is, in glacial time, nothing more than a blip.

Inside, the mates track the *Palmer*'s progress on a paper map. According to their faint pencil marks, we are currently on top of the glacier's tongue. Back in 1991, when the chart was printed, this entire area was frozen solid, a part of Thwaites's Eastern Ice Shelf that extended nearly a hundred miles farther out into the Amundsen Sea. Rick calls me over to the navigational console to look at the *Palmer*'s electronic course-plotting system. In the image on the screen water appears blue, the ice shelf gray, the place where we sail white and "unnamed." That's because it was, for thousands of years and literally up until just a few weeks ago, completely impenetrable.

"Maybe we should send in a petition to name the channel feeding warm water beneath Thwaites 'Larter Trough,'" says Basṭien to Rob as we sail over a particularly deep section of seafloor.

Rob does not respond to Bastien's joke. He's too busy watching the monitor that displays the bathymetry the Knudsen logs in real time. "It's over a thousand meters deep," he says. "Deeper than the gravity inversion models predicted." Just like that, on our very first morning, we make a discovery: that more warm water is likely working its way under the glacier than we thought.

"God, this bit must have just broken off in the last week or so," Ali says when a canyon appears, running between a big tabular and the rest of the glacier. A brief discussion as to whether we ought to sail around or through ensues.

"That's too tight," Captain Brandon says as he stares into the freshly cut chasm. "There'd be no way for us to outrun a tidal wave should calving occur."

I sprint to the galley, scarf down two hard-boiled eggs and half a cinnamon bun, and run back up, taking the stairs two at a time. Soon I am outside again, attempting to honor the ice by looking away as little as possible. Thwaites's calving edge stretches over a hundred miles, and so it takes us hours to travel its length. Sometimes the margin appears steep and sturdy and sheer; in other places it loses its sheen, seems chalky and distressed. We turn a corner and the face rockets upward into a wall. A wild line twists along the top of the shelf, tracing gorges into the blue-white snow. Then, just as abruptly, the parapet has crumbled, cluttering the water with floating pieces of brash.

Gui and Joee join me out on the bridge wings. She's wearing green Carhartts and a black thermal shirt, while he's bundled up in a puffy winter jacket, a wool scarf, and a beanie. Of all the people onboard, they are the two with whom I feel closest, and yet I don't think we three have ever spoken together alone. For a while, we whisper about the fog clinging in dense pools where the ice meets the water, and also about how underdressed the woman from Maine appears compared to the guy who studied whales in northeastern Brazil.

"It seems fitting to me that Thwaites is unwilling to yield her secrets, that everything is shrouded in mist," Joee says. "At this basic level, we really don't know what we're looking at."

"This morning, as I took pictures, I felt a little like I was—I don't know—stealing," I say. "Like, who am I to take a picture of this place?"

"That big block looks like it could fall any second," Gui says. Then he purses his lips and blows in a mock attempt to make it tumble.

"Maybe if we holler at it," Joee suggests.

We let out a series of yips and growls, each of which returns as an echo hollowed out by the ice. We are running close now, with only a hundred yards or so between us and Thwaites. The sea still. Its mercurial surface reflecting the glacier's damaged face. Our whoops turn to laughter and then silence. Because nothing we do or say matches what stares back at us, the deceptive calm we can't digest.

An emperor penguin dives off a nearby floe. Begins porpoising in and out of the water. It draws closer, turning, I imagine, to look at the ship. What about us registers as novel to it? Our size? Our speed? The sounds and smells we emit? The white of the bird's underbelly flashes teal through the Southern Ocean's stained glass. Then the graceful creature pulls away from the bow and, after a few long seconds, soars out of the water. The flame on its long neck a stamp of light.

Tasha comes over. "Try not to take any photos with the smokestack in them," she says. Behind her a tendril of gray vapor unfurls in the otherwise cotton-colored sky.

"That can't possibly be from us," Gui says. He squints into the distance. "Oh yes," he adds, shakes his head. "Yes, it is."

Tasha leaves, and we do what she asked us not to.

What I wouldn't give for this glacier not to be breaking, for the expedition not to be here at all. But here is where we have arrived, after so much effort. Our elation matched only by our grief.

BECKY:

My father says my birth was really special for him because my mom had to go into surgery afterward. It wasn't a medically necessary surgery—it was totally fine—but because of that, he got to hold me and spend more time with me right after I was born. He really loved that moment, I guess. He gave me a little white dog with pink ribbons on the ears and named her Francine.

When I went away to college, I left Francine at my mom's house. I got in my Blazer and I drove up to University of Kansas and not three weeks later Katrina hit, and my old school, my home, my entire life was underwater. I couldn't get hold of anyone because they had all been displaced. That stuffed dog was the one thing that when I went back into the house—actually one of two things—I kept. I said, *I'm not going to leave this behind.* She was up in the attic, so she was okay. A lot of things in the attic weren't as moldy. But still, Francine was pretty nasty and gross. I washed her like a million times.

I had wanted to be a paleontologist when I started school, to study dinosaurs, and in particular large marine reptiles and their migration patterns, but Katrina and all the work my mom did to bring our neighborhood back left an impact on me. I started talking with professors about global warming. I got really excited about it, because of my personal tie-in, my coming from New Orleans and all. I could make the connection: science to impact.

> It's hard to say just where or when exactly my journey to Thwaites began. Does it have to do with my being from New Orleans or living through an unprecedented storm? Maybe. But it's also not that straightforward. What I do know is we've worked so hard to get here and we've waited so long. We also know that in the past, sea levels have risen really quickly. But what's different is that, this time around, we're causing it. We're part of the biggest science experiment ever conducted. Right now, we just keep throwing our poker chips in and hoping we don't lose it all.

DURING THE SAME FALL THAT I chat with Meghan Kallman and Diana Richards, the same fall that Hurricane Dorian causes over a billion dollars in damage and record high temperatures occur all around the United States, the same fall I finally read *Casting Deep Shade*, C. D. Wright's poetic exploration of beech trees (one of the many species to have flourished in Antarctica when the world was considerably warmer than it is today); during that same fall, Felipe and I go camping at a lake an hour west of Providence. We try just before we pile into the car and take off. And again the next morning in our little orange tent. He cooks breakfast over the campfire while I do a shoulder stand on the spongy forest floor, hoping that gravity will help.

The following week, I join millions of others in the largest climate protest in history. Together we walk out of our schools and workplaces to call attention to the fact that the very structure of our world is also our undoing. I wear a dress covered in green gladiolas and blue sandals. Park my Jetta by the train station and head toward the city center carrying the pizza-box sign I made the night before. On the corner of Canal Street, a student from my creative nonfiction class holds up a yellow poster with the words "WE DESERVE A FUTURE" stenciled in big black letters. We speak about his essay in progress and also of the day's unseasonable heat.

By the time I run into Will and Chloe, members of the environmental activist organization I joined upon returning to Providence, I'm sweating. We walk together toward the statehouse and discuss the letter we're drafting to Governor Raimondo, asking that she repeal her nomination for public utilities commissioner because the candidate is a former lawyer for the gas company, National Grid. "That position, it's one the appointee can hold for life, so they'll likely shape Rhode Island's energy policy for years and years to come," Chloe says. I'm still new to the group and grateful for the background information. When we reach Burnside Park, where a critical mass has begun to form, Chloe chuckles. "We're, like, the oldest people here, by a lot," she says.

I stand on the lip of the stone fountain to get a better look at the crowd. Most of those gathered can't be much older than twenty. They hold up signs that say "I'm sure the dinosaurs thought they had time too" and "THIS IS AN EMERGENCY." Three young women in Sunrise Movement T-shirts take to the podium to talk about the ways climate change is deepening inequalities: the heat waves that hound the vulnerable in the underserved corners of our cities; the floods that dismantle low-lying, low-income communities first; the many species pushed to the edge of extinction by a system that fails to value the diversity on which we all depend. And then, one by one, they pivot, talking about how the fight for environmental justice can lead to collective liberation. Their eloquence and their ability to recognize what it has taken people of my generation years to learn—that the climate crisis is both a threat and an opportunity multiplier—floods me with adrenaline.

Evan Travis, a fourteen-year-old from South County, grabs the microphone and says, "These individual actions we're taking, like turning off a light or recycling a plastic bottle, are barely making a dent in the crisis!" He talks about how he didn't know what to do anymore, about how he couldn't sleep at night because he was so nervous about the future. "But on March 15th, I attended my first climate strike and learned more in those two hours than I had my entire life before. I learned about community and how, coming together, we can actually make a change." The crowd goes wild. He adjusts his grip on the teal smartphone that holds his speech. Someone yells, "When our planet is under attack, what do we do?" The response resonates in every direction: "Stand up, fight back."

Inspired by the chanting students, I lift my sign a little higher. It reads "Seas are rising and so are WE." Last night, on the kitchen floor, I surrounded the word *we* with jagged lines—I wanted it to appear explosive. As I pressed the orange marker against the oil-stained cardboard, I thought about how that *we* included everyone at the rally and the millions organizing worldwide, and perhaps also the blastocyst I hoped was implanting in my uterine walls, the spark that I want so deeply to nourish and protect. As I look out over the sea of singing students, for the first time in a long time, my awe is not tempered by exhaustion, disgust, or fear. I do not try to question it. Instead I embrace the sensation of possibility, of germination, that radiates throughout this growing body.

BASTIEN:

I know that my mother, her waters broke the night before I was born, and she spent a solid twenty hours getting rid of me. She definitely went for an epidural. I was the first. And yet it was still a slow and painful process for her. My little sister once asked for details in all their gruesome glory. There was some tearing and some suturing and a lot of shit everywhere. It was like many births, I suppose.

My great-grandmother died the day before I was born. I'm not going to say, outwardly, that I think I might have a little of her in me, because that kind of sentiment is very unscientific. So I will just quietly nod, and let you draw your own conclusions.

My family was very prepared. My dad planned the route and drove it daily for a week just to make sure he knew

> the way. They lived deep in the countryside then—well, not *that* deep, but it was a little way to the hospital. There was some flooding, and they had to change the route at the last minute. But they had the suitcase—everything, really—ready. We Questes tend to overthink, overprepare, overstress.

HOURS LATER, THE *PALMER* MOTORS ten miles from the ice front, out into the center of the nameless bay, to where Bastien plans to deploy the first glider. The afternoon remains foggy and oddly serene. Puddles dapple the forest-green deck. He and Joee carry the hundred-pound contraption out of the Baltic Room in a simple mesh stretcher.

"The reason I do this is for the toys," Bastien says. "Some people have classic cars they like to rebuild. I just really like cool oceanographic instruments." His first cruise, back when he was a grad student, was canceled, so he spent nearly a year overhauling his advisor's gliders. He rebuilt all of the data-processing tools and sensors, started thinking of the fleet as his own. "I guess if you call it a fleet, that makes us admirals," he adds with a goofy grin.

Each of the machines in the fleet is named after a different species of whale. There is Humpback and Omura, Minke and Orca, and this one, Melonhead. Together they lay the glider down near the crane tower. It may look like a five-foot-long yellow warhead, but it doesn't have propellers or an internal engine. Instead, a pump subtly changes the craft's buoyancy, which causes it to sink or rise. A couple lithium batteries power the relatively simple process. Meanwhile, the big side fins, or wings, create lift, helping it to glide through the water column. The objective: to generate a profile of what is happening from the surface all the way down to the seafloor through a series of tidal cycles.

Joee radios to the bridge for a winch operator. Then she cinches a lasso beneath the machine's fins and attaches it to the A-frame's pulley system. A big berg slides past. Somewhere in Great Britain, someone is preparing to send mission commands to Melonhead because our internet connection is not strong enough to reliably reach the hundred-thousand-dollar machine.

"I can't even manage a tweet's worth of text," Bastien laments. He touches the radio in the breast pocket of his orange life vest, as if he might be able to communicate with the glider that way. Then he runs his hands down the length of his dark-blue jeans and turns his gaze back to the sea.

"We didn't even know that circumpolar deep water (CDW) was working its way onto the continental shelf until very recently," Lars told me during transit. "The first time someone tossed a sensor overboard and saw something like this, they said, 'Oh fuck, we're doomed.' It really was a revelation." We've known for about three decades that Antarctica's great glaciers are losing mass. That the sea, not the atmosphere, is driving their untimely demise—this information is more recent and more troubling. That's because water has a much higher heat capacity than air. Taken together, all the CDW in the Southern Ocean holds 1.2 quintillion kilojoules of energy, or the equivalent of warming the entire atmosphere in the same region to 400°C. Which is why, as this band of dense, salty water works its way under the ice, solid slips back to liquid and the shelf fractures. But the temperature of CDW is not uniform from top to bottom, and it can change depending on any number of factors, such as the weather or the angle of the moon. Meltwater flowing from the continent can mingle with the warm stuff working its way in, causing it to cool somewhat. "It's like when you pour a little milk into your tea but on a much larger scale," Lars says. These subtle shifts can make a big difference: the distance between two degrees is, after all, the difference between something frozen and something fluid, the difference between ice shelf and ocean.

For a few moments, the ship's engines stop. A thin band of grease ice floats on the water, holding the surface almost still. Joee arranges the excess rope, hand to hand, in long loops. She lays the bundle down, adjusts her hard hat, and asks, "Are you ready?"

Bastien tightens the screws on Melonhead one last time. He stands and nods.

Then Joee lifts her hand high above her head and pinches her fingers together. The crane turns on, the glider lurches from its cradle. Its wings onyx, its body yellow and shiny as a new taxicab.

Bastien pats it on the belly and says, “Don’t fuck up.”

Five bergs gather on the bruised horizon. The crane maneuvers the glider out over the ocean and lowers it down.

“It’ll be gone in two minutes,” Bastien says, nodding, confident as always.

Melonhead’s body begins to sink. The wing tips disappear. Just before the orange transmission antenna slips beneath the surface of the sea, I close my eyes, blow the glider a kiss, and make a wish: that Melonhead returns with lots of information, so that we might understand what we’ve seen today—from the cracks in the ice shelf to the bergs in the bay—as well as everything that passed before us, but which we had no way of noticing or naming.

Part Two | *Nameless Bay*

SETTING: Never before has a human being set eyes on any of this. When the author says as much to her shipmates, they repeat the same thing back, but still, somehow, it does not seem real. She tries not to ask them how they feel about being here because she suspects that they, like her, haven't got it quite sorted yet. Are they discovering someplace new or drowning in it? Nothing about the last continent is clear, though she thought the ice would sharpen her understanding of whether they stand at the end of the world or the start of a new one.

The week after the climate protest, I attend a presentation titled "The Deluge" by oceanographer Baylor Fox-Kemper at the Brown Faculty Club. The day is hot and the dining room is packed with fellow professors. In addition to being an avid cyclist and running a lab of his own, Baylor is a coördinating lead author of the "Ocean, Cryosphere, and Sea Level Change" chapter of the IPCC's Sixth Assessment Report. He synthesizes information like that which the glider gathered at Thwaites, to provide policymakers with the science necessary to drive decision making. I like him immediately: his nasal voice; his hip black glasses and silver earrings; how he gives a shout-out to his bandmates, who are also in the audience; and the way all of this purposefully plays against the image of the expert projected by his twill jacket.

"One of the things that's challenging in terms of talking about climate change is that it can be a very depressing topic," he opens. "So I'm going to try not to sound flip, while still making this presentation interesting." The slide behind him shows a sixteenth-century engraving of an inundated country town. Some people are drowning, while others construct scaffolding to raise them above the waves. A woman with water up to her waist carries a walking stick in one hand and a baby in the other. He tells us that sea level rise can feel almost biblical in its scope and suddenness, and so we fear it will sweep us away.

Then he starts talking about the deluge of changing information and new insights that impact our complex climate models. "Computers are getting exponentially faster every year, that's true. But small physical processes will have a direct impact on the earth's operating system—things like what's happening at the boundary layer, where ocean and ice sheet meet—and those very specific processes won't be resolved into our global climate models until 2100 or maybe even 2200." Baylor talks about the coming physical deluge: how even if we stop pumping greenhouse gases into the atmosphere today, seas will keep rising for centuries. He talks about the geological deluge: that ice sheets and glaciers can take hundreds, even thousands of years, to respond to environmental changes, which means we have only just begun to witness Antarctica's transformation. He also talks about the deluge of data points we swim in: that the number of papers written about these processes has increased exponentially over the last couple of decades, and yet emissions levels continue to rise too. He talks about the deluge of little decisions we face every day, and how each is, in its own way, a distraction. "Plastic bags actually take less carbon to produce than paper, but it's silly for us to get all caught up in figuring out which bags we ought to ask for at the store because there are experts who know the answer, and their knowledge ought to drive our policy making." I slice my grilled chicken breast and spread a ball of butter on a dinner roll.

Toward the end of the talk, Baylor starts to run out of time.

He wants to discuss the deluge of responses to climate change, from disinformation to adaptation, defiance to humility, but as he starts to skip slides, the story line falls apart. "We're going to lose some of the things we most value, and until we have the full emotional space around that, we're also going to have a hard time making sensible decisions," he concludes.

In the question and answer period, a professor asks for clarification. "Are you saying that policy making is more important than individual actions?"

Baylor pauses, then responds, "That's not quite right. Collectively, our individual contributions will be the solution to the problem. But individuals don't have enough information—and probably never will—to know what we ought to do. So science and policy have to fill that gap. We can target the biggest problems and work on them first. But the actual working on them comes back to coordinated individual actions." He finishes by saying that the climate protests of the past week might mark the kind of cultural shift needed to effect large-scale political and economic reforms down the line. "Young people have more at stake. They know they'll experience directly an awful lot of change in their lives," Baylor says.

All afternoon, I ponder the phrase "coordinated individual actions." I can't tell if it is an evasion—gesturing at community while maintaining the familiar foundational principle of our liberal democracy, that the solitary human self is the bedrock of modern society—or if Baylor is suggesting something subtler and more profound. Perhaps the line between individual and collective action is not nearly as static as I've been imagining; perhaps they're reciprocal, with one giving way to the other, then flowing back again.

The following afternoon, in my advanced seminar, the students discuss an essay about the biases programmed into facial-recognition software. I listen but am preoccupied by my breasts, which feel tender and expansive, testing the stretch of my cotton tunic top. By this point, I know that desire molds my mind, makes it good at playing tricks. We try and try again. I have imagined being pregnant many times, and many times I have been wrong. As amber light filters into the classroom, I wonder whether I will look back at this moment. It could be the start of radical change—or it could be the beginning of another round of disappointment.

The next morning, while frying eggs in the cast-iron pan, I lift my shirt and turn toward my husband.

"Do my breasts look bigger to you?" I ask.

"Yes," Felipe says, "yes."

Three days later: two pink lines.

SALAR:

Time doesn't mean anything anymore, other than it tells us when we're getting up and when we're falling asleep. But that doesn't even really count 'cause it's all morphing into one long day, right?

PETER:

My "day" starts at a stupid time: I get up at eleven at night, having tried to get to bed by three in the afternoon. And then I have midnight rations for breakfast. But that's not breakfast food. It's always burgers or beef or pasta or some stir-fried pork: you know, something you wouldn't expect to have half an hour after you wake up and brush your teeth.

BASTIEN:

Yesterday we made a joke about that Friday feeling, so it must be Saturday? But days of the week really don't mean anything anymore. It's not like we get donuts on Sundays or anything.

RICK:

My routine has to stay steady. I always want to have my wits about me when I'm on watch, so I've got to stay healthy. I take my vitamin C, try to stay away from colds and flus. I exercise every day, which keeps my energy levels up. And I don't skip on the rest. I get my two to three hours in the afternoon, and I get my solid six in the night, and I don't deviate from that.

BASTIEN:

I've had eight hours of sleep over the last three days.

BECKY:

Last night—well, really, it was yesterday morning—I had so much mud in my hair, I had to take a shower before lying down.

GEORGE:

When you're on night shift, you sleep during the day, but it's not deep and uninterrupted. You're always kind of semisleeping. It's not nice, biologically scheduled sleep.

JACK:

I've been in this mode: just cook, eat, sleep, wake up, repeat. I'm not completely going insane, though we have our moments. I keep telling Julian the next time we're in here together, I'm gonna stick a GoPro on my head, and we're gonna start a YouTube channel.

ALI:

I don't have downtime, I just work the whole day, and then I go to sleep, and then I get up and work the whole day. Some people show up right before their shift starts, but I'm more of a—basically, I'm just here. Becky's the same, but she's more extreme. She sleeps when there's an availability to sleep.

BECKY:

I always think about the autotrophs, the little critters in the ocean that make food for everybody else—it's hard for them to sleep when the sun is out too.

A SWITCH FLIPS WHEN WE arrive at Thwaites. No more sauna club or bridge lessons. No more ping-pong down in the hold or king cake at midnight. All that matters is data. Mud samples, seafloor depths, temperature readings, and wave action; we even keep a real-time log of sea ice observations. Hypothetically, each of the different teams slices their days in two: some people work from noon to midnight and others from midnight to noon. But most rest only when skating dangerously close to delirium.

I too lose track. Know only vaguely that February will soon turn to March. In my bunk, I roll over and check my phone. 4:56 a.m. I descend the ladder to the floor, pull on my long johns and Carhartts. Zip up my warmest sweater and step into my XtraTufs. Open and close the door quietly. In the hallway, try to walk without making noise. On this deck and the one above, someone is always asleep, no matter the hour. Down in the galley, Ariel lays out chafing pans of bacon, eggs, and pancakes. Confused, I check the clock on the wall. It's three hours ahead of my iPhone, which must have picked up an errant satellite signal in the night and tried to get back in sync with astronomical time, with the rhythm that moves nearly everyone else on the planet. I turn off automatic updates in the phone's settings, then reset the clock so it matches the one in the mess.

When I reach the Dry Lab, I check the whiteboard's running tally of what has been accomplished and what lies ahead:

~~2:58 Kasten Core 004~~
~~3:22 Megacore 005~~
~~6:04 Megacore 006~~
~~7:46 Jumbo Gravity Core 007~~
~ 10:35 CTD

Above the whiteboard hangs a sign that reads "DO NOT USE WATER ON FIRES IN THIS AREA." Next to it is a *Calvin and Hobbes* cartoon. In the first frame, Calvin asks his father, who is working on an early desktop computer, why ice floats. "Because it's cold. Ice wants to get warm, so it goes to the top of liquids in order to be nearer to the sun," his father responds, forefinger pointed toward the ceiling in a know-it-all gesture.

"I should just look stuff up in the first place," Calvin says, turning away.

Little flourishes like this comic strip keep popping up all over the ship. Some are intended as pure entertainment and others provide advice or warnings, but those that mix registers are the funniest. Like the photo of a penguin by the bathymetry console, with the caption "What does a penguin get if it sits on the ice too long? Polaroids. (Don't check your email or look at pics on the Knudsen compy!)" James sits in front of the sign, fiddling with a Rubik's Cube. He is a solid eight hours into his shift, a significant portion of which he has spent right here, monitoring the sub-bottom profiler's rendering of the seafloor. This job is straightforward, ongoing, and of supreme importance: the person on watch must constantly match the strength of the sound wave to the distance it will travel; otherwise the image that returns will be illegible. They also log the coordinates of those locations where the drape of sediments appears thick and striated like layer cake, making it a good candidate for coring.

"The sampling seems to be going well," I say, gesturing to the list of crossed-off tasks.

"Go see for yourself," James replies. "They're working on the Kasten right now."

The aft Dry Lab is almost library quiet, save the faint whoosh of the lab fan, the squeaking of a Sharpie on plastic bags, and the rush of water filling the slop bucket in the sink. Becky hunches over the Kasten core's far end, a clipboard on the table beside her. She doesn't even look up when I enter. Instead she dabs a bit of dirt on her fingertip, lifts it to eye level, and gazes at it through the loop of her hand lens. The top of the three-meter-long, fourteen-centimeter-wide metal chamber has been removed to allow for easy access to the mud within. A couple of days earlier, when I commented on the jury-rigged appearance of the CTD rosette, Joee said, "Wait till you see the Kasten core, it looks like something built in your backyard." *Or like a miniature feed trough*, I think as I stare at the galvanized box brimming with sludge.

The procedural science at the heart of sediment sampling is simple: lower a core overboard until it hovers just above the seafloor. Then let it drop. The force of gravity, combined with the weight of the instrument, will push it into the ground. Sometimes the core comes back empty. Sometimes it traps a strange creature inside. Occasionally it hits a rock and breaks. But if all goes well, as was the case last night, it returns to the surface chock-full of mud.

Behind Becky, dirty clothes hang from a makeshift line. Hard hats litter the countertops, along with plastic spoons, kitchen gloves, carabiners, a roll of paper towels and another of tinfoil, six sponges, three peri bottles, a bunch of spackling knives, and a box of wipes. The general disarray is a sign that the night shift has been busting ass. Scrawled across the back of Becky's float coat are two words: MUD BUGS. It's a reference to both the single-cell creatures she studies and the crawdads she eats back on the bayou. Before I can ask her a question, Victoria shoots me a look that says, *I wouldn't interrupt if I were you.*

Becky and her team work with three core types, each designed to sample different slices of geologic time. Megacores, with their organ-like array of plexiglass tubes, gather as many as twelve samples, all from the top foot or so of the sediment column, offering insight into the earth's most recent rumblings. Usually the mud impounded in a Megacore was laid down sometime between the start of the Crusades and the present day. The length of the Kastens, on the other hand, combined with the weights loaded onto the top of the shaft, helps them to punch through the uppermost slurry, retrieving as much as three meters of mud. Finally, the extra-long, extra-heavy Jumbo Gravity Core (JGC) can pull in five meters or more. It offers the least precise information but can reach the furthest back in time, sometimes providing insight into what was happening all the way to the Last Glacial Maximum, roughly twenty thousand years ago—when sea levels were about four hundred feet lower and Antarctica was twice as big, her glaciers extending all the way to the edge of the continental shelf.

Currently, the Kasten looks like a human being on life support. Sticking out from the mud are ten plastic syringes, slowly siphoning off some of the water locked within. They are placed down the core at critical intervals, each propped atop an overturned red-and-white Coca-Cola cup from the galley. Like the Kasten itself, the method used to extract samples is anything but high-tech. South of 70°, the less complicated the instrument, the better; things are bound to go wrong, and there is no hardware store to scoot off to, no plumber to call. Still, each Kasten must be sampled in nearly two dozen different ways. On the far wall, another whiteboard keeps track of the team's progress. Beside each step, a box waits to be checked: Core description. Torvane measurements of the mud's strength. Pore water. Samples for biological material and trace-metal proxies, including lead, cesium, carbon, and nitrogen. Smear slides. The list goes on and on. Each tube gets picked apart in the same way, each inch of sludge put to good use. Even the leftovers get archived. The first time the team worked through this process, it took them nearly twenty hours to complete, but by the end of the cruise, they will be able to catalog a Kasten in less than ten.

Soon I hear the scratch of pencil on paper. Becky begins meticulously cataloging the color, texture, structure, and composition of each centimeter of the Kasten's contents. "Because we don't know anything about this area, we're going to learn from whatever we sample, from whatever the rocks and seafloor tell us," she says. Becky starts every description at the core's base because she wants to look at the sediment in sequential order, tracing a line from the distant past to the present day. "It's like flipping through pages in a book," she adds. Her words remind me of an idea that Julie Cruikshank explores in *Do Glaciers Listen?*, that those who live alongside glaciers tend to understand the events of their individual lives as always unfolding "within a scaffolding of much older narratives," narratives in which human beings are not the only actors. The stories told in the "books" of mud that Becky studies begin over fourteen thousand years before the earliest written records of human experience. Long before anyone pressed a reed stylus into a clay tablet, Thwaites spat sediment that speaks today of its last large-scale retreat.

Victoria holds a Munsell soil-color chart over the mud, trying to identify its exact color. "What do you think?" she asks, pointing to two different swatches. "Brown or yellowish brown?"

I squint next to her, and we decide that it is closer to the color of baby poo and thus the latter. In the corner, Kelly melts wax so she can hermetically seal the liquid extracted from the core into glass vials. The curve of her lower back flexes slightly as she stirs.

"I need the music turned off," Becky says, rubbing her eyes. A little later she will tell me, "We find in a lot of settings all around Antarctica that the last hundred years or so looks really different than what has happened over the last fifteen thousand. Like really,

really different." For the first time in fifteen thousand years, nearly all the glaciers are retreating simultaneously.

The air grows thick with the smell of asparagus pee. "The lab sink's clogged again," Cindy says as she makes her rounds, clipboard in hand. "That's why it stinks."

I follow her over to the far wall, and together we go at the drain with a plunger and a couple of gloved hands. Waves throw spume across the window, where it freezes into a creamy film. The light changes from bright to brooding blue. Eventually Becky, who has been awake since our arrival at Thwaites, announces she needs rest. "If this storm dies down, get me," she says. It's not yet ten in the morning, and already the wind is blowing hard. The ghostly forms of recently dislodged bergs slide past the portholes.

"You got any music?" Rachel asks the moment Becky leaves the room. I plug my iPhone into the homemade stereo that Barry fashioned from an old amplifier board, an AM/FM circuit, and a wooden box. When I flip the little metal toggle, a reedy Colombian *cumbia* rushes into the lab, making it feel temporarily warmer. "Music helps much of the tedious work that's so necessary to science get accomplished," Barry told me when I inquired after the improvised boom box's origin story.

"What else can I do?" I ask.

"I'm going really slow. Maybe you could help me by holding the sample bags open and rinsing the tools?" Rachel says. This is her first time in Antarctica, her first time spending months on a research vessel. When we spoke on Skype prior to departure, she came across as deeply shy, often letting her advisor answer in her stead. But over the last couple of weeks, the timid girl who stepped onto the *Palmer* in Punta has become someone else entirely: someone wide-eyed and attentive, irreverent, and seemingly at home in the world. Someone with big clumps of clay caught in her hay-colored hair.

"I'm at centimeters 140 to 142," she says, gesturing to the middle of the core.

I grab a bottle of ethanol and drag a five-gallon bucket over so I can clean the plastic spoons and spatulas between samples without having to run back and forth to the sink. Then I fish the bag labeled NBP1902 KC04 140-142 out of a stack. "Here you go," I say, holding it open over the core.

Rachel scrapes away at the mud, careful only to gather dirt from that exact location.

"Hold on," I say, as she shoves the spoon into the bottom of the bag. I gently squeeze it around the bowl, coaxing dirt away from the utensil. Twice more, and the sack is full. I fold the flap at the top, sealing the sample shut. There are a lot of things I don't know—like where the *Palmer* is exactly or where we are headed—but I do know that it feels good to work alongside my shipmates, immersed again in what I am starting to understand is a deeply communal act.

"Centimeters 142 to 144," Rachel says, hovering a clean spoon over the clay.

"Confirmed," I say. Then I read the numbers neatly written on the bag in my hand. "Centimeters 142 to 144."

Rachel nods, then digs in.

As we progress down the core, the texture and consistency of the mud changes, often in subtle ways. We know that some sections of the sediment are riddled with tiny bottom-dwelling organisms called forams, almost invisible to the naked eye. Like miniature snails, forams are surrounded by a shell. There are about seven thousand species in the world, and some are pickier than others in terms of choosing building materials for their coverings. What Becky and her team hope to find are *Bulimina aculeata*, a foram in the shape of an ice cream cone that lives in circumpolar deep water. They construct their shells out of carbonate (the same stuff that fizzles in Tums), a material that, if undissolved, will provide the team with accurate carbon dates, giving us a sense of when CDW first started working its way under Thwaites.

"Ali and I have studied so much material that's been gathered on the *Palmer* over the years," Becky says a few days later, reflecting on what the scientists have accomplished thus far. "And now we're the ones actually doing the collecting. I love that you can track the evolution of how our thinking about certain problems changes over time by looking at what people carry back." Becky's advisor, John Anderson, was among the first to analyze sediment from the Southern Ocean. Period. Back then, they simply wanted to know what was down there. They tossed cores overboard at regular intervals to create a base understanding of Antarctica's marine geology, its deep history. Today our focus is much narrower—we seek out particular sites for the information they might provide about deglaciation.

Together we work toward the core's terminus, a task that takes nearly two hours. When we reach the Kasten's snout, Kelly suggests slicing the massive sample in half with cheese wire, but before we do that I go to see if there might be bigger Ziplocs in the kitchen. Jack tells me to look around but says he doesn't know of any. I take a quick tour through the walk-in refrigerator, the shelves of dry storage, and the galley, where a big pot full of potatoes bubbles on the stove. Instead of returning empty-handed, I carry back a cup full of dried peaches to share.

After we lift the giant glob of mud into two separate sacks, we snack. Below the rumblings of the twin Caterpillar engines and Lauryn Hill crooning, "Ready or not, here I come / You can't hide," below Kelly and Rachel remarking on the delicious solitude of the night shift and their shared appreciation for Jack's high-quality birthday cakes, below all that, a piece of what was once part of Thwaites scrapes against the hull. It drags down the length of the ship with metallic pings and odd echoes, then it's gone.

Ali pops in, a sign that the day shift will soon take over. "It's blowin' a hoolie out there!" he says—which, he informs me, is slang for both a strong wind and a rip-roaring party. Then he adds, "Wow, you guys made great progress."

"Yeah," says Rachel, as she gives the metal chamber a maternal pat. "At some point I was just doing some random tasks, and I looked up and looked around and thought, *I'm in a lab on a ship in Antarctica. How cool is that?*"

"Why don't you go take a look at the Knudsen profiles so we can talk about where to core next?" says Ali.

"Are you sure you want to ask me that question?" Rachel responds. "How else are you going to learn?"

I pop a piece of dehydrated stone fruit into my mouth and chew. The nectary taste of it travels down into my chest, cracks my breastbone a little and cranks open the ribs that guard my pulpy heart. What would it mean, I wonder, to permanently take up residence here? Alongside one another and the sloughing ice. In a room lit by a shared question.

KELLY:

You don't usually get a chance to have so many people
focused on one thing. You know how it is back home.
Everyone's day is full to the brim. But on the ship—
well, there are some really good things about being on
the vessel together for a long time, just really focused.
Everything gets amplified. You get into a groove with
other people and you get a lot done.

ROB:

Last night we did a CTD and then used a variety of
coring devices near the tip of Thwaites's remnant glacier
tongue. We don't fully understand what is going on there
at the moment. The Kasten core is really good. We've got
some Megacores. Good generally. But when we finished,
the weather was not good. It's been blowing hard for the
past couple of hours.

RICK:

Well, when I got here, it was four o'clock in the morning,
the beginning of my shift, and it was really foggy.
Limited visibility and quite a bit of ice everywhere.
When I first arrive on watch, I'm always trying to get
acclimated. Captain was up, and we slowly made our way
to the ice edge. And then we started to circumnavigate

the edge of Thwaites's tongue. We were traveling along and trying to look at the mystique of it all, I suppose.

ALI:

We were a couple hundred meters in front of Thwaites. So about as close as you can get to the ice shelf front without getting under it. We have this band of sound, right, that we use to chart the seafloor. The sound also fans out away from the ship, and when you're that close to the ice, it can tell you how deep into the ocean the ice shelf front reaches. It varied a lot, but in general the ice didn't extend more than half a kilometer into the sea.

BASTIEN:

Melonhead is in the middle of the big trough, the one that feeds water into the smaller, unnamed one that we keep trying to call the "Larter Trough" just to piss Rob off.

BECKY:

It's really exciting to see things come together, to know that we're getting the material we need.

AROUND THE SAME TIME THAT I speak to Meghan Kallman, I begin in earnest to try to track down Emilio Marcos Palma, the first person born in Antarctica. In the few books and articles devoted to women on the ice, he appears with stunning regularity, but the information contained in those sources is scant and often identical. During the austral summer of 1977, a transport sloop dropped a pregnant Silvia Morella de Palma and her three children at Esperanza Base, Argentina's flagship research station on the tip of the Antarctic Peninsula. There they were reunited with Captain Jorge Palma, the army physician and officer in charge. The last two months of Silvia's pregnancy passed without incident, and on January 7, 1978, she gave birth to her son, Emilio, the first Antarctic native.

For the first half of the twentieth century, the main motive for Antarctic work was territorial conquest, with Britain, Norway, New Zealand, Australia, France, Chile, and Argentina each claiming a slice of the pie. But in 1958, in the wake of the Second World War, many of these nations, together with Russia and the United States (despite the escalating Cold War), participated in the single biggest cooperative scientific enterprise ever undertaken on the planet: the International Geophysical Year (IGY). The idea was to study the earth and its environment, with most of the attention focused on the poles, which were only recently made more accessible by technology. Tens of thousands of scientists from sixty-seven nations participated, their labor resulting in the launch of the first earth orbiting satellites, the discovery of Van Allen Radiation Belts, and the start of systematic and ongoing observations of the atmosphere and ice in Antarctica.

Intense diplomatic activity followed the IGY, culminating in a draft of the Antarctic Treaty, which set aside the continent as a place of peace and science. Military operations and weapons testing were prohibited, as was eventually the extraction of any minerals or materials of value. Instead of voiding all previous land claims, the treaty simply overlooked them for the length of its tenure, which then was a minimum of thirty years, and today extends until 2048. But the agreement didn't lead to a total cessation of the scramble for Antarctic land. While most government-run Antarctic programs prohibited sending pregnant people to the ice, Argentina purposely deployed an expecting mother with the intent of strengthening its future territorial claims after Chile's dictator, Augusto Pinochet, visited his country's Arturo Prat Base. The subsequent birth of a baby boy on the tip of the Trinity Peninsula is both part of a familiar story of nation-states vying for control of land and people and also one of the most wild and most intoxicating things I've learned about Antarctica to date.

My search for Emilio Marcos Palma turns up a couple of unexpected gems. I find five photos taken shortly after his birth. In my favorite, Emilio's older brother cradles him in his arms.

His sister stands beside them: her hair in two high pigtails, her snow pants zipped up over a sweater with a strange geometric pattern woven down the center. All three wear expressions of disbelief: hers is open-lipped wonder while the brother's tight mouth suggests skepticism; the babe, in white hand-crocheted baptism booties, appears utterly shocked, his arms up by his ears. I find an Associated Press wire story that says Argentina's president, Jorge Videla, personally sent the family a card and a handful of celebratory gifts. I discover that seven more children would be born at Esperanza. A school set up and a chaplain shipped in, too, so Argentina could also claim the first civil society on the southernmost continent.

Eventually I write to Pablo Fontana, who heads the humanities and social sciences department at the Argentine Antarctic Institute. While he doesn't have a way to get in touch with Emilio, he offers to connect me to Natalia López, who lived with her three siblings and parents at Esperanza for a full year in the nineties, when she was twelve. Like Jorge Palma, her military father had been deployed there to support scientists and was, after a yearlong tour, granted the right to bring his family over too. At first Natalia and I exchange emails and text messages. She sends me a link to the website she maintains about women working at the poles and another to her radio show on the same topic. I send her the contact information of a few female MTs who deployed with me to Thwaites. Finally, nearly half a year after I first reached out, we Skype.

Our conversation is frenetic at first, both of us finishing each other's sentences, agreeing that the familiar stories of conquest fail to depict the intimacy that the ice also fosters. "What I remember most is playing outside without danger. Or with a lot of danger from forces of nature, but it was nothing like the danger of living in the city. There was no one who was going to rob you of anything, because we lived very much *in* community, with each other. They became, over that year, my family," she says.

Natalia tells me there were almost two dozen kids at Esperanza back then, that her mother worked for Argentina's public radio, logging dispatches from the ice, and that the galley always served pizza on Saturday. "I remember eating with all the other families special Christmas food," she says. "Around midnight one of the military men came in dressed as Papa Noel. I knew he wasn't real, and I also knew that many of the other, younger children didn't know that yet. We could see him walking across the ice. It was snowing, and he carried a gigantic orange bag. He passed out gifts, the things our parents had thought to bring for us." I listen and ask more questions to coax out specific memories: her most terrifying moment, her most treasured. I am trying to get at the texture of a year, a year that probably felt like seven to her back then. She gamely keeps searching for the right words to give expression to the experience. "Antarctica was my whole world," she says, still astonished, "and it was awesome." So awesome that she has returned four times since to make a series of documentary films about quotidian life on the base.

Soon her baby wakes and starts babbling in the background. I'm not sure what our conversation has accomplished, but I am deeply grateful to this young mother for her time. Before hanging up, I ask if she happens to know how to contact Emilio. She says she will see what she can do. Two days later, she sends me a phone number. "It's a landline," she warns, "and I don't know if it works, but it was all I could find." For a while, I put off calling, excusing myself from the task because the number she provided is missing an area code. But, with a little help from Google, I finally figure out that Emilio might live in a Buenos Aires neighborhood called Villa Urquiza. It takes me a few more weeks to build up the courage to cold-call the first person to spend the first month of their life under the endless Antarctic sun. I call and call, but no one ever picks up.

A FIERCE STORM BLOWS IN with little warning. All operations stop, and the ship slows to a measly five knots. Those who haven't slept in a while, like Becky, disappear, while ribbons of wind-driven snow thrum across the dark sea. When the flakes settle, they do not melt. Instead, because saltwater has a lower freezing point than fresh, the ocean holds them in cold suspension. Viscous, pearly strips of fallen snow coalesce, some forming flotillas of ping-pong balls. Yesterday this bay was glass, pliant beneath the *Palmer*'s steel hull; today it is striated, each transverse band rich with depth and motion: a galactic belt in miniature, a universe unfolding.

"With time, those stripes will thicken into gray ice, and then the landfast ice will follow," says the captain. His gaze steady on a patch of snow-plastered sea in front of the boat, his right hand running over the folds of his brown flannel shirt. "Just wait a week or two, and you'll see." The aperture of what we are now calling Nameless Bay, which had opened only a few weeks prior, prepares to close.

"We've got gusts up to fifty knots," Rick reports.

"Barometric pressure's dropping too," Bastien says as he shuffles by in his slippers.

On the radar's glowing projection: half a dozen big icebergs sail past, just beyond the limit of what we can see. Never before have I been in such proximity to a potential source of physical peril without being able to actually trace it with my eyes. When I look out the window: water, only wild, rearing water, which makes it seem as though we are not moving at all. Later, at the midday science meeting, most people arrive weary. Every single activity proposed on the whiteboard has been crossed out, not because we completed it, but because, for now, we can't. On a photo in the galley of the *Palmer* pushing through the ice pack, someone has scrawled, in purple erasable marker, "ALL DECKS SECURE. EVEN THE BOW."

Rob walks over to the map table and says, "The sole contribution to science to be made is sonar survey, but we have to run with lots of overlap because the data keeps dropping out on one side or the other due to wave action."

The submarine team—which had been slated to deploy Ran in the afternoon—groans. The much that must be done compressed again into the increasingly little amount of time we have left. I return to my workspace in the corner, slip on my headphones, and start transcribing interviews. But something about the activity doesn't sit right. A berg, recently calved from Thwaites, slides past, and I am momentarily tempted to mourn for all that has been and will be lost. But the emotion is a dull, familiar one—the waves that accompany climate grief have been rising and falling within me for many years. For so long, really, that even the most eloquent elegies I craft sometimes sound stale.

As I stare out the window at the strange galactic assemblages suspended in the stormy sea, I consider a thought that has become increasingly persistent since we set sail. While I sometimes volunteer to help the scientists, the other two journalists onboard have maintained an emotional (if not always physical) remove from them as they work. I know this distance helps to generate objectivity, an ideal that I am blessedly less beholden to as a creative nonfiction writer, and yet the choice surprises me. How could one even imagine reporting on this community for months—taking the time and energy of its members by demanding ongoing explication and reflection—without returning the gesture, without offering to aid in the accomplishment of its aims? Witnessing how my peers' work practice shapes their interactions with their subjects has helped clarify my own relationship to what it is that I do. Lately I've started to believe that the act of writing isn't enough. Not because there are limits to what words can accomplish and who they might benefit—although this might also be true—but because there's such a small amount of time remaining in which we all collectively have to accomplish so much.

Meghan looks up from her computer screen and inches her rolling chair closer to my own. "I'm thinking of getting a three-centimeter line tattooed on my thumb so I can measure samples more easily," she confides.

I laugh and tell her it doesn't sound half bad, making the human body a more obvious tool. And with that, I click Save on my transcription, close my computer, and walk back over to the Dry Lab to continue helping the core team sort through mud.

LUKE:

I was born out at sea on a small sailboat. King Neptune was my father. [*pauses*] Naw, I'm just kidding. But I was born on Long Island. Then we moved to Connecticut when I was young. But, you know, I don't know much. I have a twin. I'm the second. My sister is one minute older than me. She shoves that in my face all the time. I also know that the power went out. Right in the middle of the birth, the hospital lost power because there was a big winter storm. I was born on the solstice, the winter solstice. So that's always a birthday story that comes out. My mom—everybody, really—was freaking out, but then the generators kicked on after a minute or two. Those must have been a long couple of minutes. I can imagine.

ALI:

I'm a twin. I was born after my brother. My mum was C-sectioned. She didn't plan it that way, but that's what they often do. Her blood pressure was bad, so it was an emergency, and I was quite premature. Two weeks, I think. I came out blue, apparently. My dad thought I was dead. It took me a while to breathe. I was in intensive care for at least a week or more, then I developed my lungs and I was fine. I developed my voice. It may surprise you, but I was a screamer.

I don't know much else, I suppose. After I was born, my dad went back to work, but the UK that year had a real bad winter, and it snowed and snowed, and my mum was snowed in with two newborn babies for weeks. Maybe I remember the first snows. Maybe that was it—maybe that's why I go to Antarctica. I spent my first weeks in a cold little cottage. Surrounded by snow.

After an afternoon of smearing starchy potato liquid onto glass slides, after rubbing dollops of mud from the Kasten core onto that sticky substance so Becky can catalog microfossil species, after letting each slide dry and then slipping them into a neatly organized storage container, after dinner and a little dawdling on the bridge, I head to the Aquarium Room, at the very back of the ship, to help sample the first Megacores retrieved from beneath the remnant ice tongue. Up until this moment, I have not had an excuse to spend much time in this part of the *Palmer.* It is primarily the MTs' territory, the home base for deploying and retrieving the various instruments that make Antarctic science possible.

The ceiling of the Aquarium Room is low, the air damp and filled with the gurgling sound of water coursing up from beneath the sturdy, crosshatched decking. There are two big white ice chests to the right; a jack, a five-gallon bucket, and a bunch of sawhorses to the left; and on the far wall, a pressure hose and a metal door with a plate that reads "CAUTION HARD HAT AREA." Our being here a sign that things are about to get very messy. I step into an orange British Antarctic Survey jumpsuit and pull on purple waterproof gloves, then join the others who have showed up to work. We make a scrappy gang, looking more like prisoners assigned to wash dishes than scientists sampling an extremely rare sediment core.

Far more delicate than its name suggests, the Megacore looks like a crystal chandelier with twelve cloches, each roughly the width of an appetizer plate. The nimble design preserves the sea surface–sediment interface and serves as a compliment to the Kasten, which, because of its weight, tends to disturb the topmost layer of mud as it drives deep into the seabed. Sometimes the Megacore makes it back to the boat with all twelve tubes full, but more often—on this cruise, at least—the total hovers closer to five. From this particular location there are three, the first ever to be retrieved from directly in front of Thwaites.

Ali sprays down the outside of the pipes, the watery mud running in ochre rivulets beneath our feet. Then he lifts the first

Megacore onto what looks like a giant spool of thread. The contraption pushes a rubber stopper up through the chamber, raising the mud, bit by bit, to the plastic's upper lip. Before we can slice off the sediment in centimeter increments and store them in sample bags, however, someone has to get rid of the band of seawater floating at the top. "Who wants to do the honors?" Ali asks, holding up a piece of tubing. "It's the same as siphoning gas."

I bend over, wrap my lips around the improvised straw, and suck. Since I have been awake and working for nearly fourteen hours, the shock of near-freezing saltwater in my mouth pumps me full of adrenaline again. I peer at the tube. The mud is luscious. The color of milk tea and pocked with pods that glisten like gasoline on wet asphalt. I reach out and lightly tap the top of the core with my gloved hand, lay that mud-covered finger on my tongue.

"What are you doing down there, weirdo?" asks Victoria.

"Oh, nothing," I respond, as a little of what the glacier left behind spreads through my mouth, smooth and silty.

"This is the archive. It'll be stored at the Antarctic Core Collection at Oregon State," Ali says as he shoves a slender cylinder through the Megacore's center. Home to over twenty-four miles of sedimentary samples, the university's Marine and Geology Repository is among the world's largest libraries of mud. As I watch Ali add a gray cap to the archive core, I promise myself that the next time I'm in Oregon I will make a pilgrimage to this sacred site. Much impresses me about Antarctic science, but perhaps nothing more than the mandate that all data gathered on the continent must be made widely available. In most disciplines, gaining access to relevant data is one of the single largest line items on a budget: either you must purchase it or produce it yourself. But thanks to the Antarctic Treaty, the data gathered here must be made freely available to all. Any well-intentioned scientist can write the Antarctic core curator and ask to withdraw a sample from what Ali has just set aside.

We begin to work in earnest. Ali uses a putty knife to slice off the uppermost section of mud. Victoria holds out a bag. I inch the bunger up.

"It's kind of like eating a Push Pop," I say.

"This bit also looks like baby poo," says Victoria. She runs her fingers along the seal on the sample bag, then sets it faceup on the countertop.

We get into a good rhythm. Meghan begins to file the samples into the ice chest in descending numerical order. I douse each dirty spatula in ethanol, then drop them into a big bucket of water. Carolyn has joined us. She swivels the shotgun mic as we progress. Soon I'm swaying from one foot to the other and humming something that sounds vaguely like "Eye of the Tiger," an instinct that kicks in only when I am either very excited or very tired or, now, both. The minutes pass in a cool, damp blur. Spray, dip, rinse, repeat. My mind contracts to fit the task. Some part of me wonders whether I ought to stop. Exhaustion, as I learned in the endless onboard trainings, fuels mistakes. But another, febrile instinct overrides it. There's nothing more satisfying than laboring alongside other people to accomplish a shared goal.

"This is more like chocolate pudding," says Meghan, ten centimeters down the core. "Or chocolate soft serve."

"Chocolate mousse," calls Victoria a little later. Each layer carries us further back in geologic time, until, after an hour, we are finally at the bottom of the tube. According to Ali, this might contain mud a couple hundred years old.

"But really, it's too soon to tell," he says as he gathers up some samples in his arms. Meghan and Carolyn follow him into the walk-in with the other two dozen or so baggies.

"Let's get a jump on the next core," I say to Victoria. Over the course of the day, my confidence has grown to the point where I feel that I, too, am a member of this unconventional society, where extraordinary effort is put forth to gather what seems like such an ordinary thing: a couple of pounds of dirt sucked up from the seafloor in an oversized straw. I walk to the storage rack, bend over the wooden stand, and lift. But I do so without remembering the black plastic plug that holds the sediment in place, forgetting that I must support it with my hand. In that moment, I forget so much of what I have been taught about the global systems of support that make life possible, the very ones that are falling apart as we work. I forget that without the support of an ice sheet, the glacier behind it accelerates, loosening more ice into the sea; forget that without the support of the sediment that the Mississippi River long delivered to the Louisiana bayou, the land slips more rapidly below the surface of rising waters. Forget that without the support of the friends and family back home, none of us would be here in the first place. I forget all about the constant attention and care this labor demands.

Suddenly forty centimeters of sediment are sitting on the floor like a massive pile of soupy dog poop.

My mouth fills with saliva, my head with rocks. The air hot with my mistake. The gurgle of water mixing with mud—washing away what was so painstakingly gathered—sounds strangely dense and overwhelming. I stare at the mess and almost vomit.

Victoria blinks. Uncharacteristically, she says nothing at all.

Ali returns and looks at the second Megacore ever to be extracted from directly in front of Thwaites disintegrating into the rivulets running beneath our feet. Then he looks back up at both of us.

"I . . . am . . . so sorry," I stutter. My face flushes. "I didn't hold the stopper in place." It is possible to calculate how many tens of thousands of dollars it cost to retrieve this dirt, but I am less worried about that than I am about the precious information it contained, now lost.

Before Ali can say anything in response, I add, "I'll tell Becky."

"It's okay. Get some rest," Ali says. "You can tell Becky in the morning."

Carolyn cringes as if to say, *You're on your own with this one.*

I nod and unzip my jumpsuit. Hang it on the line to dry and walk wordlessly up to my cabin. Fully clothed, I throw myself into my bunk, where I stare into the thick folds of its gold polyester curtain. For the first time, I wish myself away from the *Palmer*. What I wouldn't give to wake up and eat breakfast with someone who knows nothing of this glacier, its melting, or the data I destroyed. I consider rooting around at the bottom of my diminished drawer of treats, in search of the pack of cigarettes I brought along just in case. Then I consider walking down to the Iridium phone and calling my husband, who always knows just what to say, but it's the middle of the night in Bogotá and I don't want to wake him. *So this*, the chiding voice in my head says, *is what it feels like to be without some of the things you think you need.*

I breathe through my nose, counting to eight, attempting to drown my worry that the trust I painstakingly built with my shipmates will crumble once they find out. Even worse, my mistake might damage the relationships between the other reporters and the scientists. The self-righteousness that had been growing in me as I watched Carolyn and Jeff file their stories without getting involved—the feeling that my way of working was superior—takes on a toxic tint. How conceited and damaging that thinking suddenly seems. I am an outsider. One who, in attempting to contribute to this community, has instead taken something irreplaceable away.

If only I could stay hidden behind this curtain for weeks. But for the first time in my life, my responsibility to those around me is physically impossible to escape, no matter how briefly. Eventually I get out of bed and walk down to the computer lab, where I guiltily mop up the mud my team tracked in. Then I mop the hall. It's nearly midnight. Chris, the misanthropic lab manager, hands me a Jolly Rancher, which I almost hand back, so convinced am I that I do not deserve it.

VICTORIA:

I miss my family, of course.

MEGHAN:

My cats.

BECKY:

I miss the kids.

VICTORIA:

I have this random obsession with Frappuccinos—so,
you know, I miss driving to Starbucks and drinking one
alone in my car.

BECKY:

Coffee and Dr. Pepper for me, real coffee.

MEGHAN:

I miss sitting up in bed.

BECKY:

Yeah, sitting up in bed without getting a crick in your neck.

ALEKSANDRA:

I miss being alone.

GUI:

I miss my son, seeing trees. Green stuff in general. I miss drinking beer.

BASTIEN:

I did have a strong longing for beers after the glider deployment. And I've also had a strong longing for kombucha lately. I want something. I want a soft drink. I want a soft drink. I wanted a beer at the start, now I just want a soft drink. I don't care as long as it's something other than recycled water, goddamn it.

JACK:

I feel like if I had just one shot of Jack Daniel's, I would be okay, I would be good.

MARK:

The immediate thing that comes to mind is embarrassing, but I miss my van. I built it this summer. It's a VW T4 that I converted. We go out on the weekends—my girlfriend, my dog, and I.

JULIAN:

I miss driving with my windows down. I keep 'em cracked even in the rain, even if it means getting splashed a little in the face.

JOHAN:

I miss stuff I can do just because I like it: working on my motorcycle, working on my boat, cooking. Not because I like cooking so much, but because I miss being able to decide what to eat and how it should taste that day.

BASTIEN:

I thought I would miss my little habits. My routine. The things that ground me and take my day over. But this time, I miss people. I just feel really far away.

THE MORNING I DISCOVER I am pregnant, I board a train bound for New York City. I sit on the ocean side of the car and look out over the Ragged Rock Creek Marsh Wildlife Area, the old art colony at Cos Cob, the rose-colored apartment complexes rising up in New Rochelle, the many homes they each hold. All the while I am drifting in and out of a quiet, almost erotic state, aware that my body is having a conversation with another body whose unfurling has only just begun. Together we enter the long darkness that leads to the city. We navigate through the sweaty labyrinth of Penn Station to the southbound 1 train, ride all the way to its end, walk over to Whitehall Terminal, take the ferry across the harbor, and stroll a half mile down Richmond Terrace and up the stairs to the apartment that my dearest friend, Elise, and her husband rent. An apartment that Felipe lived in before they did. I drop my bag by the door and sit down on one of the Craigslist chairs she recently repainted to catch my breath. *Our breath?* I wonder.

"Want a beer?" Elise asks. She opens the yellowing refrigerator.

"As much as I'd like one," I start to say, but she knows before I finish. Immediately, she is wrapping her long arms around me. It feels like no time has passed at all since my last visit, where we chatted about the possibility of growing feet and collarbones and clavicles; it feels like it always does when returning to an old home, as though nothing and everything has changed. This time, the conjuring has fallen away. What we had imagined just *is*. The cells of my child are busy arranging themselves into a brain and nervous system and spine. I sip bubbly water and laugh while one being becomes two.

After mopping the computer lab, I head to midnight rations in the galley. At the table closest to the buffet, the THOR team discusses their progress. This debrief over what is breakfast for half the folks seated is how most "days" now start. I plop down on one of the stools and poke at the bacon on my plate. Victoria talks about the smear slides. Ali begins to go over the headway made on the Megacores.

"I have some bad news," I break in. "I lost one."

Kelly, James, Rachel, and Tasha all stop eating and stare.

"One of the Megacores," I hear myself elaborate.

For a while no one says anything at all.

"There will be others," gently offers Ali, with whom the news isn't so new. The generosity of the gesture is not lost on me. But after a month of transit, weather delays, and medical evacuations I'm not as confident as he.

That's when I notice Rachel—quiet Rachel, who has become so much more acerbic and confident since we set out. The corners of her mouth droop. She gets up to clear her plate.

"That core," Ali says, his voice dropped to a whisper. "We told her it might be the backbone of her dissertation work."

I don't know how to express my shame at having undermined Rachel's work, everyone's work, everyone's stake in this work and its outcomes. This is the flip side of collaboration: individual actions can diminish collective progress as much as they can contribute to it; the relationship works both ways.

"I'm sorry," I say to Rachel when she returns. The words sound feeble on my tongue.

"What's done is done," she responds, refusing to raise her gaze to mine.

"This is Antarctica," Kelly adds. "We all make mistakes."

A forced smile rearranges my face.

"No one was hurt. That's what matters most. And we still have enough mud for a complete data set. It's not like you dropped the only one." Ali follows his consolation with a tale about two former colleagues who had to break the news to their advisor that they had lost the team's sole thirty-foot-long Jumbo Gravity Core. "At the very last second, each chose to blame the other," he says with a snort. "So that strategy really worked out." Tasha reveals that she lost a Megacore in East Antarctica a couple of years prior, and Meghan mentions misfiling vital GPS data. But none of their confessions dissolve the guilt lodged in my gut. Eventually I stand, drop my dish in the rinse sink, and return to my rack.

BASTIEN:

If you fuck up, you lose ten percent of your science, but if you moan about it, then you lose twenty. I don't think Antarctic scientists are so special. We're just put in situations where we have harsh deadlines and we're given the resources—for example, four meals a day cooked for us—so we can pull our fingers out of our ass and actually get some shit done. It's stressful, sure. But it's the keeping upbeat for two months straight that's the difficult bit. That's where practice comes into it.

ANNA:

All of the people onboard are positive people, which has made it okay to fail. That's not always the case, let me tell you. Often there is a person or a group of people you don't want to know you fucked up because then they're going to gloat. That's because often academia promotes competitiveness, not collaboration. But here, it is okay to say, *I fucked up really big time*, and everyone will laugh and say, *Yeah, join the gang.* Then someone will say, *What about this, have you tried that?*

ALI:

On a vessel like this, with so much going on, things happen where a piece of kit fails and breaks, people make errors. And you have no choice: you just have to clean up the mess and keep going.

GUI:

If you spend a lot of time in the field, you learn it's better to be positive, even when things are not perfect. Otherwise you just—I used to be like this—you just are very intense all the time, and suffering a lot, and you wear yourself out. And worse, you wear out those around you.

MEGHAN:

Once you decide you're going to be accountable for your actions, then you can live with them. The funny thing about doing fieldwork in a remote location is that you really don't have a choice.

THE NEXT MORNING, I FINALLY work up the courage to tell Becky, the person who has the most invested in the cores and to whom I am most nervous to confess. She's down in the Dry Lab with her back to the door, the word STEMINIST stamped across the navy cotton of her long-sleeved shirt. Judging by her posture and her concentration, she will soon begin cataloging the next Kasten, which came onboard in the night. I desperately don't want to interrupt her, perched as she is on the edge of the activity she has sacrificed so much to do.

"Becky?" I start.

She turns and looks up from her clipboard, her attention as polished as a piece of obsidian.

I tell her everything, carefully. I end by saying, "I know how much this data means to you, but also to all of us, which is why . . ." I stop short, still looking for the right words to express my regret.

Silence.

"Well," she finally says. "Now we know you won't do that ever again." Her blue eyes momentarily soften under the fluorescent light, and I feel a surge of relief.

"For sure," I reply, grateful and surprised.

As much as I want to flee the Dry Lab, I force myself to stay. We work another core, moving again from the snout up. But, gun-shy, I don't dig in the mud; instead I sit at the end of an adjacent lab table, labeling hundreds of sample bags in penance. At some point, Meghan plops down next to me, her green neoprene jacket zipped all the way up to her chin. The smell of permanent marker cuts the air. Together we fall into an easy silence, our bodies and minds bent toward the completion of a simple yet essential task.

"Can you two do me a favor?" Becky asks a little later. "Fill those Coca-Cola cups with soapy water, so Victoria can sieve through the sediments and start separating out the forams."

Meghan looks at me and smiles.

"Of course," we both say.

Part Three | *Underneath*

SETTING: The bay rearranged, catchall for what the storm broke. All this debris. This gray morass. Moved, undoubtedly moved. Draw closer to the confusion, the ice shelf carved like the edges of a key, the doors it once held shut unlocking.

Many hours later, nearing night—after Peter runs dozens of water samples through the salinometer and Bastien deploys another glider—Anna walks from one side of the bridge to the other, taking in the sea ice conditions. The sky and the Amundsen are both a deep shade of shale, the glacier that separates them two-toned: the face gray as granite, the domed top pearly like quartz. Mist as thick as smoke. Just two—or was it three?—days ago, this part of the ice front appeared solid and sheer. Now, after the storm, it's craven, slumping. Islands of flotsam spin in this bay without a name, the space between us and Thwaites cluttered with brash and other broken bits of ice.

"Do you like it?" the mate on watch asks Anna.

"Yeah, so far. But it's moving," she responds. "We've already updated the flight plan, I'd like to not do it again."

Despite the brooding light and shifting sea state, Anna appears at peace. Her shoulders relaxed, her command unquestionable. The glimmer of a smile plays across her lips as she talks to the folks down in the Dry Lab in Swedish.

"They should be done making the adjustments soon," she reports.

For this mission, Ran will "fly" about fifty meters above the bottom, creating a more richly detailed map of the seafloor in front of Thwaites. The bathymetric maps we've been able to generate from the *Palmer* are like mere impressionistic paintings compared to the near-photographic renderings Ran can record. If we are lucky, these renderings will capture the corrugation ridges that Ali obsesses over, the ones he thinks could tell us whether the ice shelf has ever before entered a period of accelerated and unstoppable collapse.

That is, if Ran returns. The extremely cold temperatures could compromise the uncrewed sub's machinery; she could crash into an unexpected iceberg or pinnacle rising out of the seafloor; the calibration of the navigational code could be off and she could resurface under a tabular or, worse, under the ice sheet itself. Stuck in a slab of frozen water over half a kilometer thick. And then there are all the things that could go wrong that no one can foresee. In which case, the $3.6 million submersible won't come back.

Anna takes five long strides from the laptop to the farthest tip of the bridge, where she can get a good view of the winch. Aleksandra, the postdoc on the project, calls in, reporting that the updated navigational instructions loaded.

"She's learning a lot," I say, gesturing to the radio.

"These cruises are as important for them as they are for us," Anna says. "It's a chance to see what working in Antarctica is all about." Anna is serious about the things that matter and playful about those that don't or that she can't control. I've admired it about her from the start. When her shipping boxes were misplaced on their way to Punta, she worried as much about the lost Nutella and licorice stash as she did about the AUV. In that same conversation, she complained about all the paperwork she was made to sign in order to sail, including an agreement not to get pregnant in Antarctica. "I mean, why doesn't it say you're not allowed to *make* someone pregnant?" she asked.

As Anna waits for the latest update, she paces and pulls her fingers through her tangled ponytail, teasing the matted hair apart. Like the memory of a dream that returns in the middle of the day, shame seeps through my stomach walls unexpectedly. Brings bile to my throat. I examine Anna to see if she is looking at me differently. To try to tell if she knows about my mistake. But she just stares at the slow-motion snowfall, so light it doesn't leave a trace on the water.

The radio crackles to life. "Bridge, this is Back Deck. We're ready to deploy."

Anna consults with Johan and Aleksandra one last time. Then she turns to the mate and nods.

"Roger that. Begin deployment," he says.

The cables connecting the crane to the sub pull taut, lifting Ran from her cradle. Inside the bridge, the radiator hums and pops while the sub sways heavy in the air. The snow increases suddenly, as if on cue. Fat flakes descend like ash settling from a bonfire. Then the A-frame tilts away from the ship, slowly lowering the AUV toward the deep gray sea.

Anna turns away from the stern. "I feel like a parent sending their kid to college," she says. "I can't bear to watch her go." I'm always teasing Anna about Ran, comparing the sub to a wayward teen—often late, tough to talk to, and prone to erratic behavior—but today it is Anna who treats the AUV like kin.

Just off the back of the deck, Ran bobs precipitously in the *Palmer*'s propeller wash. Johan, attached to the ship through a rudimentary belay system, leans over the side with a metal pole. He struggles to unclip the last carabiner connecting the sub to the ship. My breath catches as a wave pushes Ran closer to the spot where, beneath the surface, eight massive metal blades whir, holding the *Palmer* in place. Johan uses the rod to ease the sub away from the stern. Then he is at it again, long crook extended, groping at the spring-loaded gate.

Finally, a small benediction. "All right, Bridge, Ran is away."

"Nudge us up a boat length, to give her space to dive," Anna requests.

The thrusters turn on, and the distance between us and the sub grows.

Ran's propeller spins, sending water into the air. For a moment she cruises forward, then her nose pushes down, pulling the rest of her bright sleek body into the dreary sea. Anna looks up to catch one last glimpse of that giant orange jelly bean before it disappears completely.

"If things go well, does this mean you might actually send her beneath the ice next time?" I ask. From the start, gathering data from under the shelf has been the cruise's loftiest goal, but, given all our setbacks, it is unclear whether or not it will be achieved.

"We'll see," Anna demurs. Then she adds, "Ran is on her way," the words barely audible, as she lifts her hand to wave goodbye. The sea ice, like a waning crescent moon, closes around the sub's last sputter.

ANNA:

My mother turned seventy the day before I left. We went out, the whole family. My daughters, their cousins, my sisters, everyone: we took Mother out for a full day with a meal and everything. I was careful, leading up to departure, to do small things like that and not schedule extra at work.

My mother is not worried about me anymore. She's used to it. She's proud of it. She doesn't worry. She thinks it's fun. She used to worry, but now she doesn't. She knows I'm coming back.

When my mother turned sixty, I was in Antarctica. I couldn't be there for her birthday, but then we brought together the whole ship and we called her on the satellite phone and everyone was singing for her in Swedish and she loved that. She's still telling that story. When she turned seventy, she told the story of that phone call at dinner.

I have two daughters, and this is my sixth time to Antarctica. I went quite regularly as they grew up. In the past, I used to leave one letter for them to open every week. To work in Antarctica meant I was often gone for Christmas. I remember one year where I had hidden the Christmas presents all over the house and I got to talk to them on the phone and told them where they were and I listened while they opened each one.

> When I started out, there was no one to look up to. Oceanography is very male-dominated, and in the polar regions it's worse. But that is changing. Just our being here helps young women imagine a future in polar science. My daughters, they're proud of me, of it, of what I choose to do with my life. But it was hard for them when they were small. I realize that now. They paid some of the price for it. The youngest one is nineteen and the oldest is twenty-one, so I can say they turned out okay. Nowadays, they argue about who should have my flat when I'm gone.

Sweet pea baby, strawberry baby, lime baby. The weekly emails charting the growth of the fetus compare it to a fruit or vegetable. They lend linear logic to this fun house of bodily experience—one day bone tired, the next wide awake at midnight; nauseous, then not. Time stretches out, days take on new depth. Each carries some surreal marker of growth and change. One morning I wake to learn that my child's intestines are migrating from the umbilical cord (something I inexplicably think of as *mine*) into *its* belly.

New rituals emerge. At two in the morning, I often plod down to the kitchen for a snack. While I wait for my toast to pop, I lift my shirt and stare at my shifting silhouette in the window's watery mirror. In my life I've crafted lots of different things—dinners, books; I even built a wooden boat once—but this time the creative process is different. This time, I *am* the process. As I chew the spongy sourdough, I marvel that my body knows how to construct what my mind can't even begin to explain. And more. And more.

During that pregnant fall, everything starts to bleed together: myself and my baby, protest and production, radical will and scientific discovery. I tape my nipples to keep them from chafing and paint signs for another climate demonstration with the top button of my pants undone. I research day care options and write op-eds in local newspapers asking that the governor link Rhode Island's renewable energy push to environmental justice. I interview midwives and the climate scientist Michael Oppenheimer.

We talk about the summers he spent swimming off Black Rock on Block Island and how fundamentally our understanding of Antarctica has changed over the last two decades. "In 1998, *Nature* published a paper that explored whether the West Antarctic Ice Sheet was stable. Now we're talking about the possibility of it having already entered a period of unstoppable collapse," he says. We talk about how much he hates the question of whether or not to have kids, how it misses the point. "I'd already been researching climate for a decade when I had my daughter," he adds with a cluck. Which I understand as him saying, *I know how bad it could get, and I know what human actions make it worse, and bringing a child into this world doesn't change that math.* "If anything," he adds, "it just made me work harder."

In October, a friend forwards my husband a real estate listing. Though we are not really on the market, we look. The big orange bungalow is nice, but it's the two-hundred-year-old beech tree that casts a spell. The enormous trunk looks like the leg of an old elephant. I think of the beech-like fossils Robert Falcon Scott attempted to cart back with him from Beardmore Glacier, and how some say his unwillingness to ditch the weighty rock is supposedly what did him (and the rest of his crew) in. While I don't know the truth about that, I do know that on the day we tour the house thousands of hard, spiny burrs blanket the side yard. This, we are told, is a mast year, the beech seeding in exceptional quantities—which, of course, I take as a sign.

Felipe walks toward the massive tree while I pick up a casing and clutch it in my closed fist, humbled by the way its nuts guarantee both the beech's continuation and the continuation of so many creatures in the neighborhood. I'm almost envious of the tree—the way its reproduction doesn't trip the wire we've tied to existential threat—but, more, I admire its generosity, its wordless willingness to care for what is not in any immediate sense its own.

The day we make an offer on the house is also the day I dig out the leftover meclizine tablets, the ones my ob-gyn said were for seasickness and morning sickness both. The next afternoon, finally feeling well enough to exercise, I go for a swim at the local YMCA. The farther I get from the wall, the richer the chlorinated blue becomes, carrying me back in time to that first day at Thwaites, to the cracks crossing its calving face.

I slow my stroke, lift my mouth to breathe, then dive back down into this surreal echo of the glacier's going. Before setting sail, Valentine told me that working in Antarctica would change the way I see the world. For the first time since my return, this feels true, in an unforced way. I let go of a few bubbles of air, hear them burble at the surface. Sink deeper down, until my toes touch the cold tile. The cobalt kind of hums. I can't tell anymore where I end and the rest begins: Antarctica, home, Thwaites's transformations, my own, my son, and self; all morph into one large unstable whole. We dwell there at the bottom for five seconds, maybe ten, then push off and rise back up through that reminder of something monumental, how radiant it appeared even when coming undone.

"How'd you sleep?" I ask Anna, who, seventeen hours later, is back on the bridge overseeing Ran's retrieval.

"We realized quickly that the navigation was a little bit off at first, so I stayed up until four or five in the morning, until we were able to speak to her directly and change the code," Anna says. Her still-damp ponytail drips down the back of her teal T-shirt, and her white cheeks are wind-burnt from the hours she spent preparing Ran on the back deck the day before. She looks a little mangy, but so do we all. When Lars showed up at lunch one day wearing a clean red sweater none of us had seen before, someone told him to change back into his smelly seal clothes. "You're making me jealous," they said.

"The ice is moving really fast," Luke says to the second mate, who has come on deck extra early in preparation for shift change. For days we've received no satellite data, so far gone are we. Next to all of the activities listed on the whiteboard, a simple note reads "Check sea ice conditions." Luke nudges the *Palmer*'s steel hull through a turning landscape: lily pads and pancake ice pirouette and plow into larger drifts, the pockets of near-black water between them expanding and contracting like lungs.

"All the ice gathered here," Anna says, surveying the sea's white skein. "And that," she adds, gesturing to the place where Ran has been programmed to rise, "that looks like land."

If this is the same spot we left behind just half a day ago, I can't tell. I can't tell if these massive changes are a normal part of winter coming, or what normal even is anymore, since it is absolutely not normal for any of this to be navigable in the first place. A dozen massive domed bergs line up right in front of us like horses heading into a charge. Off the stern, an MT lowers a transducer into the sea. They're trying to communicate directly with Ran to override the mission code, since a tabular the size of Anna's hometown of Gothenburg now covers the proposed retrieval site.

"I can tell you one thing: this ice isn't going anywhere in the next twenty minutes," Luke says, which is of course when Ran should return.

The tech lowers the machine deeper, searching for a signal.

"In Sweden we call this bit that's close to the ship *drivis* ice. It can be a real menace," Anna says, then sighs.

Her team scrambles to come up with a new recovery position. One that might remain open water for at least the next half hour. Soon they've got some coordinates, information which is finally, mercifully, transmitted to the sub.

"Hopefully the ice won't be a problem over there," Luke says, punching the new heading into the navigational console.

At Rothera, Carl Robinson, head of the airborne survey, had shown all of the media folks the first detailed aerial footage ever shot of Thwaites's grounding zone. I marveled at the logistical feats that simply producing the video had demanded—moving hundreds of drums of fuel overland between Byrd Station and the new Lower Thwaites Glacier camp; building a snow runway and flying a turbo prop-plane to it from Rothera; the filming itself, as Carl zipped back and forth above Thwaites, creating a survey of the nearly ten thousand kilometers that comprise the glacier—but I had no idea what to make of the images, no way to assess the significance of what I was seeing.

"Can I bring it back to the ship, to share with the others?" I asked. "And maybe if we get a strong enough satellite signal at some point, I can also post it on Twitter or something?" After nearly a month at sea, the prospect of finally having something Thwaites-related to share was intoxicating. But when I dug around in my bag for a thumb drive, I came up empty. Jeff Goodell handed over his. "Make sure you include the music credits," Carl cautioned. Then we were clomping down the wooden stairs and back out into the blinding light.

The next afternoon, after the midday science meeting, we held an impromptu film screening.

In the opening shot, a bright red plane sits on a slab of wind-blown ice. In the next, the Twin Otter's propellers spin slow circles. Then it has lifted into the sky, flying away over a wash of New Age music. The perspective shifts again, and instead of looking at the plane we are suddenly inside it, gazing out. The ice shelf swims into view. Deep gorges scour the surface: each line a vein, running toward the Amundsen. At first I think: *Grand Canyon*. Then: *No, it's not like that at all*. These crevasses pile up one after another in eerie succession. The next image is even more unsettling. The day bright, the sky parched like old hydrangeas. But as the shot pans to the left, the neat order of the shelf gives way to an undulating whitish blue; it looks like someone put a bunch of plastic toys in the microwave. Then the shot cuts to technicians wearing aviation headsets and fussing with laptops. And, just like that, the film is over.

"Let's watch it again," Bastien said, and we did.

After the second viewing, it was quiet. Finally, Lars spoke.

"When you look at the latest Landsat images, the Western Ice Shelf looks like reptile skin—kind of scaly, full of ripples," he said. "But they seem so small printed on A4 paper. Now I can see those scales are actually huge icebergs with these giant gaps in between. That the whole shelf is breaking apart." He paused for a minute, shook his head, and added, "Oh, bloody hell." Then he stood and drifted back to work.

Now, as we try to track down the sub a little more than a week later, I wonder if we are currently sailing through some of that disintegration. The captain rarely leaves his perch, his eyes trained on the rotting rim of the world. The radar sweeps its six-kilometer radius. The neon display a buckshot-torn target, each pinpoint of green light a berg. I undog the door by the Ice Tower and step outside. Murmur of snow on snow, like the noises a northern forest makes when the sun comes out after a storm and what was held in cold suspension begins to drop from heavy boughs. The softest edges of the floe crumbling as they push up against each other.

We nose deeper in, carving a path through the sea's milky cataract.

White of dove and river pearl, spackle and baking soda, plaster of Paris and spent cinders. Even the orange lifeboats tethered to the sides of the *Palmer* appear wrapped in crisp shells the color of elephant ivory. I lean over the railing for a while, content to watch the pools of onyx water opening up around the boat, however briefly, as we work our way toward the new coordinates.

By the time we finally arrive at the second proposed recovery site, it, too, is covered over.

Most everyone tumbles down to the mess for lunch. New coordinates are produced. Becky and I settle into the side-by-side captain's chairs, where she explains why Thwaites's big tabulars appear relatively uniform in size. It is the first time I've seen her sit down in days. She patiently draws a diagram in her yellow field notebook, supporting her explanation of ductile deformation with a cartoon of calving ice.

Anna looks down at the paper, back up at me, and laughs.

"You really are one of us now," she says.

A gigantic smile cracks across my face. If Anna knows about my fucking up, she doesn't hold it against me. Delighted, I clear my throat to say something, but then the radio clipped to her back pocket sputters. Anna listens intently and speed-walks to the windows on the starboard side. "Ran took one of our commands and came up thirteen hundred meters from here," she says over her shoulder. "Now we've got to find her." Rob runs to his room to retrieve his personal set of binoculars. The rest of us use the pairs that belong to the *Palmer.* I spin the sights until the image comes into focus. It is difficult to distinguish the shelf from the drift, the seam between the two nothing more than an ashen slit. Amid all this thickening winter, we search together for something distinctly not of this place.

"I've got it!" Rob shouts before too long, lifting his right hand to the horizon, his left holding the binoculars tight. There, at two o'clock, in a patch of black water, a dramatic gash of orange—the only other man-made object for thousands of miles. It's hard to tell whether Ran or the drift is moving, but soon ice rings her in like a rim around a crater.

Filip, Johan, and Joee load themselves into the inflatable and scuttle across the sea. They draw close. The Zodiac stops. Johan stands and holds his arms above his head, a giant *O* for *okay.*

Anna yelps, then mirrors the goofy, triumphant gesture. "We've done it here," she declares. "We actually took a little tour under Thwaites!"

"What?" I say, blinking back disbelief.

Thwaites Glacier is comprised of three continuous parts: a floating ice shelf, the grounding line, and the land-based ice behind it. The grounding line serves as a kind of pivot point. All the ice inland of it rests atop solid earth; all the ice on the other side floats in the Amundsen Sea. The original plan was for Ran to dive to the bottom of the ocean and shuttle back and forth right in front of the floating part, cataloging one of the troughs through which CDW travels. The data she gathered would have augmented what was collected on our very first morning here, adding detail to the topographical map that we generated.

Instead, what Anna is telling me—and what I can't quite believe—is that the sub slipped beneath the shelf and swam under the ice for hours, collecting all kinds of previously unknown information. Information about the temperature of the water, its density, and the way it circulates; information about the shape of the ocean bed but also of the underside of the ice itself. There are physical processes not included in even the most localized models of Thwaites's retreat, simply because there is no observational data from which to extrapolate them. What has just occurred will change that. It's the difference between taking photos of Mars and landing on it.

Out on the water, Filip slips two pieces of bright yellow webbing under the submersible's belly, then threads them through the metal rings that run along Ran's back. He cinches her tight to the Zodiac's gunwale. When he's done, Joee stands, tiller in hand, to steer them back to the ship.

"I've learned that sometimes it's a good idea to be sneaky," Anna explains. Had she announced the mission's objectives in advance, she would have been peppered with hundreds of questions before the release, turning her precious attention outward. And since she didn't share her ambitions, I have an almost private audience to witness this historic moment. My head feels light and bubbly, like it just might pop off my shoulders and float away. Out in the water, the weight of the sub makes the Zodiac list to one side. Joee's body leans the other way. Slowly they draw closer to the *Palmer*, until the little gray craft begins to disappear into the stern's long shadow. I laugh.

I can't resist asking, "How do you feel?"

"I don't know how to describe it. I just wish that everyone gets to feel this feeling once in their lifetime. It's a very good feeling," Anna says. Her response is instantaneous and unmediated and has nothing to do with claiming this prize as her own. I tell her how honored I am that I get to share this moment with her, then I wrap her in a hug. It is one of the first times we touch and one of the last. Later, when we are all back on solid earth, Anna and I will lean against one another like gunslingers, smoking cigarettes under the streetlights of Punta Arenas, drunk for the first time in months. But for now, everything is cold and bright, and Anna walks down the five flights of stairs to the mudroom, where she puts on her black float coat and white hard hat. On the back deck, she unites with the rest of her team. Once they get Ran onboard and check over the instruments, Salar picks up a fistful of snow, wads it into a ball, and throws it at Anna. Then Filip follows suit. Suddenly everyone is tossing snow at everyone else and screaming.

ALEKSANDRA:

We just explored a new place on earth that no one has ever been before. It's not just the continent, or the ocean, but we actually got under the ice shelf. So that is—well, it's amazing.

ANNA:

It's an achievement.

ALEKSANDRA:

She's the first woman to send an AUV under an ice shelf.

ANNA:

And you're the second.

ALEKSANDRA:

In the lead-up to deployment, I wasn't even asking what's happening. I was just taking orders.

ANNA:

I like being the first under Thwaites. Not the first women. Just the first. Full stop.

FILIP:

At yesterday's meeting, at the end of it, you said, *By the way, this stays between us.* And I felt like, *Okay, let's do it.*

ANNA:

I think it was just yesterday—I mean, that I told my team about the plotting.

JOHAN:

I wasn't . . . worried. We had the bathymetry from the ship that looked a couple hundred meters sideways, under the ice. We knew the area wasn't too shallow, that there was an opening.

ALEKSANDRA:

Right, that's why you were so calm.

ANNA:

Tomorrow we will start to work through the data. Tomorrow.

FILIP:

I'm going to the sauna.

ALEKSANDRA:

I'm going to row three miles, then go to bed.

JOHAN:

Sleep. Just sleep.

ANNA:

I've got a very nice toffee sandwich to eat.

THE NEXT MORNING, I ASK Scott if I can borrow his Insanity Max 30 DVD. When his sister-in-law bought the suite of workout videos for him for Christmas, I bet she had no idea how popular they would become on the ship. But while lots of people have tried them, Scott is the only one who has really committed to the "60-day total-body conditioning program." "Busting my butt is keeping me from snapping," Scott says as he hands me the sleek carrying case. Then he pats his belly and adds, "I'm doing the workouts, but not the recommended diet. I figure I can fast on whatever healthy stuff remains when we cross back over. I don't know, canned asparagus and pineapples."

When I get to the gym, Fernando is wiping down the all-in-one trainer in the center of the room. An old home video of a housewarming party plays on the television. The camera moves from one person to the next, all smiling easily. There's a roast pig and an elderly man dozing upright in a chair, his head inching toward his chest, then snapping to attention. A bright white living room opens into a bar, where a younger Fernando laughs with a friend. He's wearing a white T-shirt with the sleeves cut off and sipping a San Miguel.

"Do you want to use the TV?" Fernando asks, seeing the DVDs.

"No, you go ahead, I'll ride the bike instead," I say. I walk over to the corner, crank up the resistance, and put in my headphones. Juan Gabriel starts crooning "Querida." On the screen, Fernando's wife sits on their brand-new couch and breastfeeds their baby. She looks deep into the camera and grins, and though I know nothing about her—including how she feels about her husband's work, which likely paid for this new home—I am tempted toward at least one conclusion: that she is genuinely happy about the arrival of this child. The track in my earbuds changes. Freddie Mercury sings about traveling at the speed of light. Inspired, I increase my rotation, stand, and start trying to climb an imaginary hill.

Fernando looks at me and laughs, then turns off the television and leaves.

FERNANDO:

This is a good story. My father and my mother had one baby. Then they started looking for a second. Because he's a carpenter, my father built a small house. Then he got some money and he expanded. The old people—we've got so many superstitious beliefs in Calape—they say this bigger house is not good. They say, *You built the bigger house, but the small house is still in the middle of it. You cannot do that.* My father didn't listen; he said these superstitious beliefs are only for the old people.

But it is true that in the new house, the bigger one, they were always sick. My father went to the hospital, then my mother is going over there too. Swap, swap, him for her, all the time. The old guy from up the mountain, the *albularyo* [medicine man], told my father, *If you don't get out of that house, you will both die.* They were almost dead just from going back and forth to the hospital. And they were running out of money. In the end, they moved just six hundred meters away, into a smaller house again.

Once they leave the big house, they get healthy and we are born, we are four, all boys. My mother and my father did not expect us. I was the first to come in almost twenty years. When I was born, they were surprised. Then my father said that the *albularyo* was right and that maybe some of the old ways might make some sense.

In the Philippines, most say, *I want a boy to carry my family name.* That's true, but I've got only one boy and now he studies in the seminary, so he's not going to carry my name. But I don't want him to change the plan of his study. I told him, *Choose what you want. Go for it. I will support you.* He asked, *What is your suggestion for me?* I

> wanted him to be an engineer. If you're working in the Philippines—or even working abroad—as an engineer, you can make good money.
>
> But I did not say that. Instead I just told him not to become a seaman because it is not easy, a life like I did. In the night I pray, *God give me a sign*. The next day, a seminary went to his school and he chose that. He will not do what I do, and that, at least, will be better.

THE AFTERNOON AFTER RAN COMES back, Joee asks me to help her on the helo deck. She's wearing a blue down jacket, patched with two overlapping hearts to keep stuffing from falling out where the shell is ripped. Outside the sun shines hard off the sea, each wave tip a tight bright diamond. I'm glad for the excuse to stand in the cold and hold a rope. The crane operator turns on the winch, lifting the Zodiac up off the deck, and I guide it over to some supports. Together Joee and I scrape ice from its gunwales, then I go back inside to the Dry Lab, where Ali sits beside Jonas, discussing what the sub saw beneath the shelf.

On one monitor, there's a rudimentary map showing Ran's flight path; on the other, a black-and-white rendering of the seafloor.

"Take a look at this," Jonas says as he zooms in on what looks like a series of tire tracks.

Ali blinks and draws closer to the screen. He stares at the image, then turns toward the Swedish technician to check to see if this is some kind of prank. Jonas takes a big sip of coffee from his travel mug and places it carefully back on the countertop.

"Someone call up to Rob's room, now," Ali says. "He needs to see these." Then he slaps his leg and emits his seal-bark laugh. "They're corrugation ridges, those ridiculous features I've been scratching my head over for years!"

The grainy monochrome image makes me think of the Apollo

landings, how they finally brought into focus a part of the universe that humans had obsessed over but never seen up close. When I say as much, Anna reminds me that we actually know more about what's happening on the moon than we do about what's under the floating parts of Thwaites.

After Ran's recovery, the mood on the ship shifted again, this time from anxious overdrive toward something slightly more sustainable. Still focused, but less deadly serious. Despite all the setbacks—the rudder trouble and medevac—the raw information necessary to better understand this place is making its way onboard in bulk. Bastien's deployed two gliders, we've dropped the CTD rosette over the side a couple dozen times, and the seals Lars and Gui tagged are already sending us information about the temperature and salinity of the water column. Scott and Meghan gathered dozens of old penguin bones, and the coring team has many tubes of Antarctic mud packed away in the walk-in. We've even mapped hundreds of kilometers of uncharted seafloor. To celebrate, earlier in the day Anna appeared on the back deck wearing a brand-new, fake-fur hat with earflaps and the logo of a Russian hockey team.

"Are you a fan?" I asked.

"A friend gave it to me because they thought it would be fun to sneak it onto an American vessel," she said, beaming. Soon, most nights, before going to sleep, I head up to the bridge, partly to take in the incremental freezing over of the ocean and partly to shoot the shit with Luke. He tells me that he had tickets to go see *The Price Is Right* with his father, but because of the delay he had to give them up; he also tells me about the houseboat he and his wife like to rent when he's home. It's the first time, in years really, that making idle conversation doesn't result in my chest going tight, doesn't produce a trapped, panicky feeling that I should be doing something more productive with my time.

Rob arrives in the Dry Lab and I step aside.

"I can see them without my glasses on!" he declares. "What's the spacing between them, peak to peak? Five meters? Maybe six?"

"Six-point-seven meters between each one, and this one is twenty-two meters wide. So far they're really regular," says Jonas.

"And there are more?"

"Yeah, they're all over. Perhaps more than a hundred different instances," says Anna.

"Well, that adds a new dimension to the story, doesn't it?" Rob asks drolly.

Ali covers his mouth with his hand and leans even closer to the screen, as though proximity will help him decipher just what put those ridges on the seafloor. I try to connect what I have learned about glacial collapse to these little heaps carved into the fine silt just this side of Thwaites's calving face. The thoughts I have I don't share, fearful that they will disrupt perhaps the most important onboard conversation yet.

"They're so uniform it's hard to imagine an iceberg had anything to do with them. An iceberg would create irregular spacing," Anna says.

"They could be tidal," Rob responds, "as in an iceberg keel lifting up and down—bouncing, really—along the seafloor."

"We've got to drop a Kasten," Ali adds. Then he looks up at the other two. "Do they all occur in flat places?"

Jonas clicks on a couple of keys to clarify the topography. "It looks like a meter of relief over the whole area, and you'll notice the material is really soft on top," he says. Barry arrives. As does Filip. Lars and Victoria and Carolyn and Jeff. Tasha too. After over a month at sea, any new activity is bound to draw a crowd, but this is, to put it tersely, why we are here: to observe what no one else ever has, to gain insight into the physical processes that are causing the widest glacier in the world to lose six times as much ice annually as it did three decades ago.

"So, a one-point-five-meter Kasten," Ali says, turning the focus back to next steps. "Just to see what they're made of."

"We wouldn't know if we were coring peak or trough."

Someone suggests sending down a camera.

"Think of the ridges at scale. If you stood atop one, it'd still be hard to tell where you are; they're actually very smooth undulations," Rob says, and the idea instantly dies.

The conversation splinters, each group groping for a way forward. The handwritten warning on the bridge, that the ice dictates our movements and not the other way around, floats into my mind. It occurs to me then that this discovery is dependent upon what Thwaites allowed. How she held her form long enough for us to pull close on that first morning, to glimpse just a little of what rests underneath.

"How full circle," I start to say, "that Ali's data—"

But Ali doesn't let me finish. "It's not my data. The whole THOR team, together we're responsible for it."

"None of this would have been possible without the bathymetry that THOR collected on that first day," Anna says. "Those maps made us confident enough to send Ran under the ice."

What we often think of as a solitary pursuit—the production of scientific knowledge—is more like an extended relay race, in which the baton just gets handed off again and again and again. The arc of any breakthrough is long and relies upon extraordinary human effort, including that in service of the most ordinary things—Fernando replacing the *Palmer*'s flooring, the captain and mates steering the ship, Jack and Julian baking tuna steaks in the galley. Whatever we might learn about Thwaites from the images Ran carried back is dependent upon the marine technicians and the people who wrote the code to run a sub in water so cold that it would be frozen if it were fresh, dependent upon all the partners back home who are taking kids to soccer practice and making dinner or running to McDonald's, dependent on the workers who are paid minimum wage to flip the burgers that feed those families.

Jonas zooms back out and selects another swath on the map. A fresh set of treads comes into focus.

"Only a few people have worked on these features so far, and they're all using data with a much, much lower resolution. There's no real consensus as to what causes the ridges and what they might mean. It's one of those puzzles," says Ali. Never before has a geophysicist partnered with an oceanographer to create this kind of rendering of the seafloor so close to Thwaites's grounding line. "I have a hunch these ridges are going to be important. Like, *really* important."

Rob claps him on the back. "I've seen them. They're exceptional. And now I'm going to go eat lunch," he says. "We've got a full program ahead of us."

JOEE:

At the end of my shift a day or two ago, we were doing some multibeam survey, covering a lot of ground, and I was up in the Ice Tower. It was dusky, and everything was that really intense blue with a little pink in the sky. We were coming along this big tabular, this big wall of ice, and when we get to the end of it, we come around the corner and this panorama just opens up right in front of me. I had this feeling that the continent was revealing some part of itself in that moment. It was so . . . intimate.

ALI:

I've spent fully half my life studying this place. So I suppose I've come to—I think of it as a really close friend, one that I see regularly. For me, there's a strange familiarity to it all. Sometimes I take Antarctica for granted because I see it so often. But other times, it's just so clear that it's this amazing force, shaping our lives.

SCOTT:

I get cheesy when I talk about glaciers. I try to make people interested by telling them that glaciers are alive, at least to me. If you sit next to one for a long time, maybe you don't see it flowing, but you eventually will see it break or calve. You will hear the water, you will hear ice cracking everywhere. When I camp out next to a glacier, all night, I hear the sounds of it, the popping, pulsing sounds of it. It's hard not to think of a glacier as alive.

LARS:

I don't know if it's gratitude I feel for ice. I don't know if that goes too far inward toward belief or toward thinking that the ice is made for me. Ice was not made for me and my enjoyment. However, I do think things like fire and ice and water, they have their own life, and when you watch them for hours, or days—which is easy—they're always changing depending on your mood, on the light, on the thing itself. In that context I have gratitude, because it's fantastic to watch what the ice creates.

ALI:

I just took a job at University of South Florida, I'm moving somewhere in the world where this, what we are seeing right now, will absolutely dictate where my wife and I buy a house, for instance. Antarctica is so far removed from my home in terms of geography, but they are absolutely connected, right next to each other, in my mind.

THE WINTER AFTER MY RETURN, I travel all over the country to talk about my book *Rising* and what we can learn from coastal communities already responding to higher tides and stronger storms. At the Cary Institute of Ecosystem Studies in upstate New York, during the Q and A, a man in a khaki vest says we are like frogs sitting in a pot of water that is slowly being brought to a boil—that because we fail to perceive the subtle changes in temperature, we will soon be cooked to death. It is a metaphor I have heard many times before. He concludes by suggesting that in adapting to climate change, we are accepting what we ought to fight against.

I stand under the hot stage lights, my pleated dress exposing my pregnant belly, thinking about how this tired binary—pitting mitigation against adaptation—is often peddled by the people for whom the difference between 2°C and 2.1°C of warming doesn't really matter. I think about those who left Paradise, California, the day it burned; about those fleeing devastating drought in the Middle East; about all the people who have already surrendered one definition of home for another and, hopefully, for the promise of safety. I think about the glacier's grounding line, how it, too, is on the move, and how no amount of present-day mitigation will fuse it back to the land it left. When I set out, I feared I would return convinced that having a child was unwise or unjust. And while that has not been the outcome, I realize in the dark auditorium that I am still trying to understand what the ice reveals by coming undone, what its disintegration asks of us.

Before I speak, I think about how I must look to someone who has spent decades making dozens of sacrifices—perhaps going vegan, driving an electric car, flying less. To him, I might look like a hypocrite, ringing all the warning bells even as I set a new fire myself.

Finally, I say something like, "It's not about choosing one or the other, and it's not about environmental sin and redemption. I'm interested in how we might be transformed, and not just for the worse, by this force that we shape and that shapes us in turn." This feels both true and wholly inadequate at the same time.

The next day, Felipe and I drive twelve miles east to the Stone Church, a cave with a cascade running through its hard center. It is February, and the local newspapers warn of brushfires come summer due to the lack of snow. As we work our way up the hill, I peel off my layers, remarking on the warm weather, on how unsettling it is for winter sun to be matched to so much birdsong in the bare limbs above. Still, the closer we draw to the cave's entrance, the more slippery the stones beneath my feet become. In the shade, ice still holds. I stop a hundred feet short of my destination, fearful of falling. Hover my hand over my stomach. Inside: a human the size of a coconut, just two open palmfuls.

I have always loved to push my limits outdoors—it brings me the peace of passing beyond the self, as well as freedom from others' opinions about how my body should look or act. In the woods, instead of being a woman, I am just another human being plodding on. But this time, I steady my swelling body against a rock and encourage Felipe to continue without us. When I look up, my husband is just about to round the corner into the cave. He pauses on the cusp, describes the beautiful chasm where the sky filters through. Then he turns around and looks at me, checking to see that I am truly all right not getting any closer.

"Go on," I say, nodding.

In premodern times, some people refused to name their children until they survived one year. In the United States, just a hundred years ago, one out of every ten babies born did not live more than twelve months. In some cities, poor hygiene and malnutrition brought that number closer to one in three. Meanwhile roughly one in one hundred mothers died in childbirth. Everyone either knew someone or had a friend who knew someone who did not survive labor. Some died from hemorrhages, others from misused forceps, many from a mysterious disease known as "childbed fever," where pus would drown the abdomen and ovaries, causing the new mother to rot away. Motherhood—which at the start of the twenty-first century, in one of the richest nations in the world, we've been taught to think of as something electable—wasn't bathed in godlight, wonderment, and pastel pajamas back then. It was a risk, one that many women had no choice but to take.

There on the hillside, I finally understand that mothers are makers of life, to be sure, but that in doing so, they are also makers of death. Not the possibility of death, but its guarantee. The bright sound of the river in winter fills the slender valley. A family of three walks by, a father holding a little girl's hand. "Be careful," he says as she heads toward the cave's icy mouth. Her pace slows, her little blue boots searching for traction on the slippery ground. Then she, too, disappears into the opening in the stone.

GEORGE:

I have very little to tell you about my birth. I know it
happened at night. My mom was a single mom, so I never

had a father. I don't have any siblings. Sometimes a family decides not to have any more children, and sometimes it's strictly a financial or economic decision. I think my birth was kind of simple: it was almost at midnight, between the 1st and 2nd of April. So when the doctor asked my mom what date to put on the record, she told him to put the 2nd, not the 1st. She didn't want me to be born on April Fools'. I don't know why people don't like this day. Maybe I like it because I was actually born on this day. I always try to make some kind of joke. I tried to make one with my father-in-law once. I made such a big joke that he was almost packing his stuff to go to the airport. But that's another story.

RACHEL:

It's terribly boring, but my parents planned the day that I was born. I was born in Colorado Springs, Colorado. The doctors told my mom she could choose between March 28th and April 1st. She chose March 28th because she didn't want to have my birthday fall on April Fools'. It all went really well. *Easy* is what she says, a C-section that she got to watch happen, which is crazy. She said she couldn't feel any pain, but she could feel the pressure change. She said it was the coolest experience, like an alien was being taken from her body. I don't know if it was the drugs they put her on, but she said that she loved it.

LARS:

I was there at my own birth, but I can't remember anything of it, as it turns out. I didn't want to be squashed—I thought being born the natural way is too much effort—so I was a Cesarean. That's what my mum says. That's the only thing. The hospital was just about three hundred meters from the ocean, by the port, in Wilhelmshaven, Germany. You just walk a few minutes and you're at the water. From that moment onward, I was always close to the sea. This is why I sometimes say I was born on the crest of a wave.

JULIAN:

My own birth? I don't remember being there when I was born. I don't remember nobody telling me anything about my birth. When I was little, I just liked to hang around the older kids. I wasn't the one who sat around on mommy and daddy's lap, listening to stories and all that crap. I was out there trying to get into big-kid stuff. So as far as birth goes, I don't remember nothing.

ACT FOUR

Part One | *The Quickening*

SETTING: Somewhere it is Sunday. Dawn breaks lambent, brilliant. For once, the dome of the atmosphere is distinguishable from ice and sea. Saffron first light fades to lapis lazuli. Grease ice pulls across the water like contrails in a clear sky, each stripe slowing the speed of the waves, holding the surface still. More bergs than ever before.

The six scheduled CTD stations have all been cancelled. On the whiteboard, a thin line runs through each set of proposed coordinates, next to the note: "ice covered." "It's fall," George says, adjusting the oversized collar on his navy sweater, a cotton one with big plastic buttons that he has been wearing for a month straight. "And down here, fall lasts two weeks, then it's winter all the way."

During the medevac, George offered his services as my personal ping-pong trainer. One of his dearest friends, another Latvian in exile, was a professional player, which is how George's game got so good. We occasionally met in the hold after lunch to play a couple of matches, but I lacked the dedication to turn his attention into anything other than a twelfth-place finish. In the days since, George and I have taken to swapping tea instead. Or at least that's the premise; it took only two servings for him to tire of my Earl Grey. Today, he brews me a cup of his best Russian black, which I carry up to the bridge to drink.

The sun bursts out from behind a flat mauve tabular, the edges sharp like splintered mica. Wind pulls spindrift off its backside in lacy lines that mirror the streaks of frazil ice in the bay. During the night, sections of the sea froze over, and a glider that Bastien hadn't heard from in almost twenty-four hours finally sent a distress signal, indicating its antenna might be encased in rime or, worse, dislodged. Now he's down on the back deck with Jennie and the MT who replaced Carmen, preparing gear for what he hopes will be a successful retrieval: gaffing hooks, rope, metal poles. First we have to locate the machine, however. Rick guides the *Palmer* through pulsing pancake ice, aiming for the origin point of the most recent ping.

Where yesterday a long line of tubulars sat on the horizon, making it difficult to draw close to the ice front, today the bergs have somehow split and multiplied. When the wind dies down, they steam. Later, I will learn that in this moment—this moment when the bergs are many, lavender and faceted; when the air is full of floating ice crystals; when I step outside and tears pour from my eyes, freezing my sunglasses to my cheeks—in this moment we are in great danger. But for now, I know nothing of the disassembling in which we float and instead think that finally, for the first time, the air feels just as cold as I expected.

I clap my hands together, then momentarily remove my gloves. Unclick the lens cover on my camera and focus on what is objectively one of the most beautiful moments of the cruise so far. The recently calved chunks of Thwaites heap up like mountain ridges. One moment the sun is a prism, sending light in every direction, then a puff of icy air passes in front of it, like chalk dust, and what was so sharp a moment ago appears more diffuse, cloaked somehow.

"It's too damn early," I declare five minutes later, when I walk back through the door by the Ice Tower.

"Minus thirty-five windchill," Rick reports. Soon we arrive at the coordinates, and everyone awake scans the dense pancake ice for the glider's signal pole.

"There it is," Rick says.

"Where?" I ask, incredulous.

"There," he says, pointing. He hands me a pair of binoculars.

"You've got to be kidding me," I say when I finally find what he is talking about. Only a foot of the neon-orange pole cants above the water line. We approach it, but the first attempt to lasso the craft is a miss. Bastien and the rest run to the stern, rearrange the ropes. Try again. This time the noose cinches tight around the body. Bastien works the knots with his bare hands, and the glider is finally brought aboard, plucked from a dense cluster of frozen lily pads, each tinged egg-yolk yellow, a sign of increased microbial activity.

I stand up top for a while longer, transfixed by the speed at which the bay transforms. Less than a week ago, this area was mostly open water, and now it is muddled by massive bergs. Winter's abrupt arrival in the Amundsen seems to have changed not just the consistency of the world we inhabit, but also its form; or perhaps something else is happening, something I can't quite imagine. I ask Aleksandra whether she can confirm that all this ice has come from Thwaites.

"Well, yes," she says, matter-of-factly.

"So that ice was part of the glacier," I repeat.

"Ice *has* to be calved from a glacier to become an iceberg. Calving is a natural process. What's not natural is the rate at which it's happening, the acceleration of that process." Then Aleksandra—who, like the rest of the scientists onboard, is very careful about where and when to link human activity to what happens in Antarctica—adds, "That's where we see our influence. In the speed of the glacier's movements."

Rick warns Luke that he's going to have a tough time getting back to the location of yesterday's sub retrieval. "They want to pick up the cNode transponder and drop a couple of cores, but I'm not sure that's going to be possible," he says, before heading over to the coffee maker to put on a fresh pot. At first Luke steers east, ducking and weaving between massive tubulars. When I leave for lunch, he's still at it; like a faded boxer dodging punches, he plods onward, trying to get us back to where we were.

TWO HOURS LATER, DOWN IN the Dry Lab, Rob hunches over his silver laptop. He's got two windows open and is clicking compulsively back and forth between them. Both contain aerial images of the study area, the first satellite information to have made its way onboard in well over a week. In one, Thwaites's western front is sturdy, a rampart running along the farthest margin of the unnamed bay. In the other, it looks as if some angry god took a hammer to the ice. Rob toggles between the two.

Cohesive shelf. Exploded lodestar. Navigable Nameless Bay, then the same inlet cluttered with a surreal confetti of bergs.

"The morning we arrived, we cruised right along the edge of the shelf," I say, looking at the first image. "It was pretty smooth, a solid wall of ice. There was some rumpling and slumping, but—"

"But over the last few days, there appears to have been a real significant release of bergs directly south of us, from Thwaites's ice front," Rob says, finishing my thought. Bewildered, he touches his dry palms to his muddy pants. In my stomach, a strange flutter, half fear, half excitement. This is also why we are here: to witness the disassembling that we previously only imagined with words, with calculations born from remote satellite images, with mathematical models. That disassembling, it appears, is unfolding right in front of us.

"We know what it was like the day we arrived, on the 26th of February. The first image was taken on the 27th," Rob continues. "And we're now six or seven days later."

"I think it's March 3rd today," I say.

"Right. So only five days later then. Sometime over the last five days, there has been a considerable change. Actually, even less than that. Because this area over here was where we deployed Ran." Rob points to the northeastern corner of Nameless Bay in the first image. We are like investigators attempting to solve a crime with just a few random stills from a security-camera feed. "So over the last two days, that is when there has been *real* significant change," he says. "It looks nearly as dramatic as the Larsen B collapse."

Rob is referring to one of the largest recorded examples of ice shelf collapse in human history. In 2002, scientists monitoring the peninsula through aerial satellite imagery watched in both amazement and horror as much of the Larsen B Ice Shelf (which is about the size of Rhode Island) fell apart over a period of two months. Prior to collapse, the tributary glaciers in the region lost roughly two to four billion metric tons of ice annually. In the years thereafter, ice made its way into the bay six to eight times faster than before, proving that when a shelf disintegrates, the glaciers it held in check *can* dump exponentially more of their mass into the ocean. Which means that in the days and years following this moment, the flow of Thwaites might also accelerate.

The folds of Rob's faded jumpsuit appear bleached in the lab's fluorescent light. The gray pouches beneath his bright blue eyes sag. He clicks from one image to the next again and makes an involuntary sound between a sigh and a grunt.

"Have you ever been on a ship where something this dramatic has happened in the area where you were working?" I finally ask. It is, after all, Rob's twentieth time in Antarctica.

"I haven't, no," he says quietly.

All of our remaining work in Nameless Bay is canceled. We will not gather cores to calibrate the corrugation-ridge data. We will not deploy another glider nor will we send Ran back beneath this ice. We will extract no more information from what might be the largest trough feeding warm water under the shelf. The less than a week we spent working along the western portion of Thwaites: that is the only time we will have. And now it is over.

Up on the bridge, the second mate listens to speed metal while steering us away from the minefield of the collapse. Eventually I step outside and turn in a full circle but rarely catch sight of the horizon line, so full is the sea with recently calved bergs. I have wanted to see a glacier calve for just about the same amount of time I have wanted to form my own family. I've held these desires alongside one another for so long, curious how one might make the other contort, to take on certain unlikely positions. In my mind, the ice would creak and groan, the ship's deck would tremble, clouds of dust would rise up into the bright blue vault of the sky, walls of water surge toward us. Bearing witness to such collapse, how could something not shift?

But this is nothing like what I expected. No cleaving cliff faces. No thunder and no echoes of rapture. I turn the circle again. To my right, an iceberg bigger than the college campus where I teach. Behind it, another, and another. Some have soft white snouts and others are glossy, their edges shining sharply in the sun.

I walk behind the smokestack to get out of the wind. The

low-grade hum of the churning diesel engine returns. There I search my memory for signs of collapse, for something, anything, dramatic. Anna updated the coordinates of Ran's retrieval a couple of times because every place she programmed into the code got covered with ice. At some point there were more bergs than before; it seemed as though they were always drawing closer. Just this morning, I asked Aleksandra about them, and yes, she confirmed that they came from Thwaites. If we had arrived a day ago, I think, I would believe that *this* was the way it was supposed to be.

The wind thrums in short guttural gusts around the hull. I have made it so far, and now I want—no, I need—to be able to recognize the glacier's movements as its own, its fracturing as a willed response to what we have done to the planet. When a glacier steps back or surges in the Arctic, those who live with the ice say it is sending a message. For as long as I have known Thwaites's name, I imagined receiving that message, that this moment of its breaking would ring through my body as warning. But I never considered the possibility that the cracks would be so large I wouldn't know they were cracks. That I wouldn't be able to distinguish berg from shelf, something whole from something broken.

Never considered that collapse could appear so still.

I can count on my fingers the number of days I have spent alongside this ice. I can count the number of hours I have been under its spell. In the funnel's familiar slipstream, I try a while longer to see the extraordinary event unfolding all around us, but whatever bolt of enlightenment I hope for does not arrive.

LUKE:

Sometimes I'd come around the corner and realize that
what I thought was the glacier is actually a whole separate
piece of ice, a piece of ice that was part of the glacier.

SCOTT:

You can see it from the air, but down here on the ship it's hard to see, right? To the human eye, these changes are not super noticeable. Which makes sense. We are very small in this giant world of ice.

LARS:

I mean, we sit on a ship with a couple of portholes. If you want to really see outside, you have to be active and go up to the bridge, so sometimes I feel like a person who thinks about the weather and looks at their smartphone instead of looking out the window. And then it's not real. I mean, the glacier is breaking apart—*yeah, yeah*—but in some ways, it's the same as if I were in my office back at the university, just with a really bad internet connection. Some part of me doesn't realize that I am here and that Thwaites is five kilometers that way.

LUKE:

Sometimes you won't see the break until you get around the back side of it. Because it's hard to see, there really aren't a whole lot of reference points. Everything can blend together . . . so it becomes quite disorienting.

KELLY:

We are surrounded by these beautiful bergs, domed on the top. In the satellite images you can measure them—a kilometer by a kilometer, big bits of ice that have broken off the front of Thwaites. But it wasn't broken off when we first got here. It just goes to show you how the system acts on thresholds. And when one is passed [*snaps*], the change is not gradual.

ROB:

There is a sadness to seeing something so spectacular waste away. Then there is something awesome about watching it happen as well. Every time you come, even if it's only once, you know this place is never going to be like it was. It's a onetime experience, every bit you see.

KELLY:

You know, we say things *move at a glacial pace*, but that metaphor doesn't necessarily hold anymore. These glaciers are no longer slow.

LARS:

When I look at the satellite image, I can see the entire area that Anna went under is gone. So yeah, *fuhhhhhh fuhhhh fuuuuuuckkkk*. I mean, this *huge piece* several kilometers across is just disintegrating. It's not a big, solid piece that just breaks off and drifts away. No, it looks more like an exploded chicken. You know: *pffffffft*!

ALEKSANDRA:

Thwaites is in pieces.

ALI:

It spooked me. To see how much ice has come out of there in such a short period of time.

KELLY:

Now we know that these events can happen and that we're tracking something real. Not some figment or invention.

During the ride to my first birth class, I admit to my husband that as my stomach grows, so too does my fear about how exactly this baby is going to travel from the inside out. The bigger the bump, the more obscene that fact seems; the more extraordinary it is that this growing child will pass through me and emerge into the world of earth, air, and moss. In *The Argonauts,* her book on queer family-making, Maggie Nelson writes, "To let the baby out, you have to be willing to go to pieces." When I read this again months after returning, it makes me think of Thwaites, and of how disorienting it was to draw close to that shattering.

"I know that I am"—I stutter. Beyond the window, leafless branches blur together—"strong and capable. But I also feel this will require something else—some loosening of control—and with that I'm less familiar."

Felipe nods, says nothing.

"Go through the light, then pull to the left," I say, gesturing to the parking lot in front of a simple Victorian. "There, by the sign."

He smirks. I bet he's thinking, *So much for loosening your control.*

We take off our shoes in the hallway, place them in a little plastic tray, and enter a cream-colored room where a recording of a bell being struck sounds from a small speaker. A humidifier runs. We are instructed to roll out a mat and to grab whatever pillows we will need to support our bodies during the three-hour class, the first of seven in a series. A handful of other pregnant couples line the walls, each reclining on little rafts fashioned from bolsters and yoga balls.

"Birth is a dynamic process," Kaeli, the instructor, begins. "One that requires both planning and flexibility, a willingness to respond to the various challenges that arise along the way. Your ability to do that depends upon the body's ability to consistently re-establish a sensation of safety." Then she slides forward, easing into a center split.

Close your eyes, she instructs.

Imagine that you have been blindfolded and told the class will end in a week, not a matter of hours. Hold hands. Walk back toward the building's entrance. Outside the wind blows. Step up. You are entering a large vehicle now, like a bus. The vehicle turns on. The road is bumpy. Eventually it stops. Get down. Then step up into an airplane. Hours later, you land on an island, where you will spend the week together. Now think about what, if anything, helped you feel safe on this journey. Then Kaeli tells us a different version of the same story. The final destination is disclosed far in advance. You are given packing lists and led through a bunch of preparatory exercises. You get on the same airplane and arrive at the same island, but this time you know where you are headed. Kaeli instructs us to open our eyes.

The weak winter light filters through the honeycomb curtains, and people shift in their seats. Only then do I realize how quiet the room has gone. When Kaeli asks for our impressions, it's almost unanimous: when we did not know where we were headed, the others say they felt unsettled.

"In that version of the story, it was like setting out on an adventure," I say. But as soon as the words escape my mouth, I realize they're no longer fully true for me. They are the words of a person who has never been on a journey that demanded extensive preparation, the words of a person who isn't carrying a baby. Motherhood is the adventure now, I tell myself, and I have already embarked.

We brainstorm a list of people, things, memories, and songs that can travel with us during labor and might help us to feel that we have what we need to survive. Mine is short: Fresh air. Felipe. Watermelon. My nightgown. It feels as if no time at all has passed. Here I am, again preparing to take a journey about which I know nothing, except that we do not attempt such undertakings alone, no matter what the stories suggest.

The second half of the class is devoted to what actually happens during labor. Often the mucus plug—a kind of protective shield around the cervix, formed during pregnancy—will pop out, signaling the start of the process. I learn that my body will release a chemical called oxytocin, which triggers the uterus to contract and causes the cervix to dilate; Kaeli tells us that if your adrenalin levels rise, your oxytocin levels may fall. "That's why lots of women stop having contractions during the trip to the hospital," she says, adjusting her teal scarf. "The body is in fight-or-flight mode, which isn't a safe state for birthing." I learn that, right now, my baby is getting all the oxygen he needs from the blood I pump through him, and that when he is born one valve in his heart will open while the one connecting his body to mine will close: all this will happen in the first ninety seconds of life. More than anything, I learn that I know very little. And that my body can do a lot without my passing it explicit instructions.

In *The Blue Jay's Dance: A Memoir of Early Motherhood*, Louise Erdrich describes the process of becoming a parent as one that is as ridiculous as it is rich, both earthbound and profound. And yet, she says, above all else, it shatters, is as close to self-erasure as humans can come. A laboring person passes dangerously close to death's dark gate in order to carry back a life, and the months thereafter are marked by a certain kind of annihilation, where ambition is eclipsed by the more immediate needs of another. All of this leads Erdrich to ask an unexpected question: "Why is no woman's labor as famous as the death of Socrates?" I am thinking of her query as Kaeli describes transition, the period late in the birthing process that marks the beginning of the baby's descent. The image of the old philosopher holding forth with a glass of hemlock in his hand, the strength and selflessness it is meant to conjure, suddenly seems not all that different from that of a person whose uterus bears down, pushing a child through the birth canal. What comes next is a series of endings that also mark the start of something extraordinary. I write this down in my notebook, though I don't know just how true it is, not yet.

My back begins to ache. I try crossing my legs and sitting up straighter, but it doesn't help. I try lying on my side, with a pillow supporting my stomach instead.

Kaeli closes by saying, "Do you want to know what separates a traumatic birth from a transformative one? They can be the same exact birth process, the same thirty-hour labor, with the same outcome—be it a natural birth, or an epidural, or a C-section. The woman who feels that she has no agency, that the process is being done to her, she will experience it as a trauma, while the woman who makes many small decisions along the way will experience it as transformation." As I roll the yoga mat back up and carry my bolster to the basket in the corner, I think about Thwaites and the hundreds of bergs she calved into the Amundsen while we were there. Transformation or trauma? Is she going to pieces to teach us a lesson or is her fracturing being forced upon her by our actions so many thousands of miles away? It is a series of questions I can't seem to settle or set aside.

JOEE:

Most of the picture I have of my own birth is imagined, based on memories of visiting my mom in the hospital after my siblings were born. I feel, before I allow myself to go down that road, I should say that my mother has never told me the story of my birth. We don't talk about a lot of these things. I feel myself shying away from it even now, like maybe it's a depth I don't feel prepared to plunge into. I think that's important to say.

My father was in the army, so I was born on a military base in the early 1980s. The military is still not really great about the whole *femininity is a thing that exists in this world* thing and I think they were probably a lot worse back then. My mom was fairly young. She had followed her high school sweetheart to West Point. She went to a historically all-female college one town over. A lot of people, including my parents, got married during graduation week in the big chapel.

I'm the first. I was born in Texas. They hadn't been there very long. I imagine my mom didn't have much community to fall back on. I think about my friends who are having children now, how they have this incredible community around them full of strong women and compassionate men. How those relationships also help

> them feather their emotional nests. I don't mean that in a derisive way at all. They're able to cultivate a healthy emotional space to go through their birth in and bring their child into. It's really beautiful, and it's one of the reasons why I hate being away from home. I hate feeling like I'm missing out on the growth of this community that matters so much to me.
>
> Anyway, I really don't think my mom had any of those things. If you've ever been in a military hospital, then you know they're extremely institutional. All of the ones I was ever in when I was a kid were this bleak, sickly yellow. They smelled bad. Women tend to have a lot of agency taken away from them when they give birth in hospitals. C-sections are pushed for. I know I was born by C-section, but that's about it. Again, this really isn't a thing that I talk about with my mom.

DURING THE EXPANDING NIGHT, THE *Palmer* sails east, aiming for what Anna calls "an iceberg graveyard," a little spit of high ground between Thwaites and Pine Island Glacier where dozens of colossal tabulars sit grounded thanks to the relatively shallow seafloor. When I wake, we're heaved to alongside one of them, the visible part of which appears twenty times the size of the *Palmer*. The meltwater coming off these bergs has both caused an algal bloom and driven up sea ice production, creating the ideal habitat for Weddell seals to hunt, rest, and molt. Lars spotted their telltale bean-shaped bodies in the satellite imagery and requested a stop. Even though a significant portion of Thwaites is in pieces, hampering our ability to gather data along the western ice front, we can still work in other corners of the Amundsen, making the most of the time we have left.

I start the day by transferring my most recent batch of interviews to my external hard drive. Type the words "Week five" on the new folder I've created, then stop. It seems as though we have been away far longer. So I rename the folder "Weeks five and six," dump the files into it, and get back to the thing that matters most: the ice, holding it in my gaze for as long as possible before it forces us to leave. Soon Tasha and I are standing next to each other on the bridge wings in the brittle cold, the temperature so low that even talking takes too much energy.

She sniffles, I sniffle; we watch Lars, Gui, Bastien, Mark, and Joee descend into the Zodiac. Where once the launch of a small craft was a significant event, today it is just another thing we do, like eating lunch or playing bridge. The seal team is not more than a couple hundred meters away from the ship when out of the stillness comes a guttural crack.

"What was that?" Tasha asks.

"Look, look to the left there!" I exclaim and point toward a puff of snow floating just above the surface of the sea. Another crack. This time we both watch as the face of the glacier momentarily turns to cottage cheese and tumbles. Beneath it, the ocean roils up into a round wave, making the seam where the ice sheet meets the bay look as though it is bending into a massive blue arch.

Tasha raps on the glass, screaming at the mate on watch, "Calving event!"

Then I am screaming too, gesturing wildly toward the crook of the bay. Out on the water, Joee, who surely heard the crack, steers the raft away from the surrounding bergs and ice floes as far as she can into the nearest patch of open water.

Over the previous weeks I'd asked my shipmates to tell me about their calving experiences. Luke recounted watching a big berg flip after dropping into the Gerlache Strait. Aleksandra explained how she calculates calving rates through satellite imagery and Bastien said he'd only ever seen them unfold on maps. Cindy showed me a short video shot on the *Palmer* a couple of years prior. She treated the one-minute clip as though it were a secret, closing the door to her office before she pressed play.

On the screen, a wave bigger than any I have ever seen rises out of the sea, its body cluttered with freshly fallen ice. "It's a good thing we got off the deck," the cameraman says as the crest spits and rumbles towards the *Palmer*. Moments later he screams, "Holy shit!" The image gets shaky as he runs away. Then, at the last second, he turns back, capturing the wave as it gallops down the deck. The explosion of spume momentarily makes everything appear white. In the aftermath, dozens of car-engine-sized chunks of ice float down the ship's length, suspended in the few feet of water that have also made their way onboard. "Oh my god," Cindy gasps off-screen. "That . . ." She trails off.

The memory of the clip, the sudden violence of it, scrambles my thoughts as I helplessly watch Joee hold the Zodiac steady in a dark strip of debris-free ocean. Two emperor penguins follow the little inflatable, diving in and out of the water, unaware, or so it seems, of the danger coming toward them. But as the wave draws closer, the troughs lengthen and the peak flattens, so that by the time it reaches the seal team it isn't more than six feet high. Joee and the rest placidly bob up and down a couple of times, then aim for a nearby floe where half a dozen seals lounge. When I look back to where the calving began, the dust has settled. The stillness returns. That I was able to notice this change at all speaks of its smallness, its scale just happening to match my own.

One of the penguins launches out of the water, attempting to join her mates on a slab of ice, but she doesn't put quite enough kick into it. She hits the edge and falls back into the sea. An act that she continues to repeat. Meanwhile the other penguins swing their long necks around and around, working out the kinks. The morning is so quiet that I can hear them call to one another, their reedy voices assembling in a chorus.

Once the tips of my fingers feel stiff and a little waxy, I retreat. "Take My Breath Away" plays on the ship stereo, followed by "Walking on Sunshine." I pull off my gloves and lay my numb hands on the radiator. Out in the bay the jagged floes and growlers kaleidoscope into a thousand different formations; penguins beat bubble contrails into the water with their flippers while wind riffles across the parts of the sea surface that aren't frozen yet, kicking up little wavelets.

The Zodiac disappears behind a berg that looks like a giant manta ray.

Later Lars disembarks on a floe only to find that the Weddell lounging there is one he has already attempted to tag—that's how dramatically the floating debris rearranges itself in the waning light. His footprints in the snow the only clue to his having been there before. I pull a set of binoculars from the bucket where they live, and scan the floes myself. But from up here, all the seals look the same, like distant pieces of driftwood.

Almost everyone onboard cycles through the bridge, pausing for minutes, sometimes hours, to stare. The sky unblemished, the sun bright. Rick and I speak about his routine in Hawaii, where he drops his son off at school and then busts out the boogie board or goes for a hike around Diamond Head; Rob spends some time comparing our mission with half a dozen others; Barry appears, fixes an irritating beeping noise, and is gone. The seal team creeps across the ice like a bunch of burglars intent on not waking the owners upstairs. There is something almost dull about it, like this is just what we do now. And yet the peace that comes from being a part of a group of people who need each other—this, too, won't hold.

LARS:

It was a really nice day, honestly. Never write about it. You can't. You can take photos, you can take videos, they will be brilliant, but I think it will still be really hard to get across what it was like to be here. Icebergs, sea ice, brash ice, ice shelves. Fantastic views, blue sky, sunshine, seals on ice floes, seals that we caught. Some of them we tagged. It is just beautiful. Very simply beautiful. We also had a tsunami, which was exciting.

GUI:

We heard someone screaming from the ship—maybe it was you. But we heard the strange noise before that, like a thunderclap.

JOEE:

I knew it wasn't thunder. And I definitely—I mean, I certainly had an adrenaline response in that moment. I felt my face get hot and I was like, *Okay, what do we do now? Where did that come from and where do we go?*

MARK:

You could see these icebergs bobbing up and then down in the distance. I figured the swell had to be big if it was moving bergs like that. We talked quickly about going back to the boat, but we realized we didn't have anywhere near enough time.

JOEE:

The bridge radioed us and they were like, *Hey, there was a calving event and a pretty good-sized swell is coming your way, you guys should probably get out of the ice.* That was exactly right. So I just put us somewhere where we had clear water all the way around. The big hazard is that one of those uneven-looking bergs capsizes.

MARK:

Everyone was staring in the direction of the calving, waiting to see what was coming. Wondering, like, *What are we in for?*

LARS:

But when the wave came to us it was just a slight swell.

JOEE:

I have spent, in the scale of your average human being, a lot of time down here. And nothing, nothing, has prepared me for what it's been like to work among all those massive tabular icebergs, all of which were recently calved. It's both pretty indescribable and also pretty hard to wrap your head around, even when you're here doing the thing.

GUI:

I feel connected to the origin of nature, of life, here. It brings me toward something very deep, as if I can see that all life started here. As if this is just part of the beginning or another beginning.

SCOTT:

Glaciers have waxed and waned for as long as they have been around. In that sense, it's normal. Think of it as the heartbeat, a kind of planetary pulse. But, on the other hand, usually some glaciers grow while others retreat. And right now, it's happening everywhere, they're shrinking almost everywhere, and fast.

ALI:

Some people say the ice sheet is dying or the glacier is dying, but I don't think that way. Glaciers go through changes, and they're in a constant cycle of birthing icebergs but also replenishing themselves at the center. This cycle will always happen, whether we're around or not, whether there are humans on the planet or not. They will go away a bit, then they will come back again. But also touch wood, because, well, Thwaites is moving so quickly these days that we really don't know what will happen next.

ALMOST A YEAR TO THE day after I departed for Thwaites, I stand in the shower, warm water running over my endlessly expanding stomach. I am thinking about the student essays I have to read, the grocery shopping that's got to get done.

Suddenly I am a bell, struck.

In obstetric circles, the term *quickening* describes the moment in pregnancy when you can first feel the baby's movements. It comes from the Old English *cwic*, which meant "to be alive" long before it meant "to move fast." There are echoes of the original significance in the phrase "to cut to the quick," meaning to damage the life-giving part of someone; it's also there in words like quicksilver and quicksand, both of which communicate a surprising animacy in what English would otherwise classify as inert.

The feeling comes again. It is startling in its suddenness. I hold my hand over the place where my child moves. This, like basically everything else these days, reminds me of Antarctica, how things we once experienced as inert are springing into action: ice sheets are splintering, glaciers shrinking, the archives they hold disassembling into the ocean that swirls around them. I think back to something Kelly said as we sailed away from Thwaites—to move at a glacial pace once signified a kind of mind-numbing slowness, but the world has fallen out of sync with the metaphor it made.

I grab the soap holder to keep from slipping.

We're near something now, but I can't quite tell what it is.

Ever since my return, I've wondered if the prolific calving we witnessed was a fecund or a fatal act, a birthing ritual or death throes? Was I witness to hundreds of icebergs being born or one small part of one large glacier's prolonged demise? Transformation or a trauma, calving or collapse? But what if it isn't one or the other?

Tap. Tap. Tap.

The baby moves again.

Both it and the ice, quickening.

Part Two | *Holding Season*

SETTING: The deep darkness before dawn returns. During that holding season, the ice forms something new every night, another layer laid down. From grease to gray to yearling, until what was translucent appears opaque.

Wake early and ascend the five flights to the bridge, heavy with sleep and exhaustion. Push open the door. Rick stands beside the captain, who turns to see about the noise. He smiles, slightly, then settles back into his chair and the endless labor of looking out. Neither says a word. The night was clearly cold and calm, and in its grip the grease ice thickened into gray. No whitecaps before us now, no twitching wavelets. What was liquid settles into a different, slower state. As the *Palmer*'s bow plows a path through the freshly frozen field, little bits of ice break off. Like fireworks unfurling, the frozen chips skitter away from the ship, their whiteness bright against the gray.

"We've been looking all night for a hole," the captain says. "For the next submarine deployment. But we haven't found one, so we're just going to have to make one ourselves."

Rick takes a big sip of coffee from his mug, sets it down, then turns the ship around to face the way we came. The trail we blazed unspools behind us like a black snake. "It won't be long before you can bust out the skates and go for a spin."

"Totally within safety protocols," I say.

"That's why we're going to let Luke go first."

It seems like nearly a year has passed since I last saw soil and trees, last smelled heat rising from the sun-warmed earth. Last night in the mess, I discovered a jar of lemon curd, which I greedily spooned on top of a bowl of vanilla ice cream. The sweetness glossy in my mouth. Now we stare as the stone-fruit sky turns to corn silk and opal rose. It takes nearly an hour for the sub to be deployed and for the sun to rise. When it finally does: a crack of tangerine across the cosmos. I look back to the path we cleared, but the ice has already eaten it up.

ROB:

We're in an extremely fluid situation. Well, if you look outside, it's not that fluid at all, actually. There's been a modification to the plan because of the rapidity with which things are icing up. We're going to grab the glider at the soonest; Bastien is just waiting for it to call him. We have some coring to do this afternoon and a set of four CTD casts to calibrate the sub. It went into the water at half past six this morning, so we can pick it up after dinner. That will max us out, I think.

The time has come to move over to Pine Island. But things could start to ice up over there just as they are icing up here. First of all, it's cold. And we have a very fresh surface layer because of all the melt. Plus the wind is coming off the glacier. All those things we expect to be similar at Pine Island, which is why we should go there sooner rather than later.

LUKE:

People really want to get these last bits of data. There's more pressure now because they realize the ice is closing in.

ROB:

Also remember to take some photos of Pine Island Glacier. It's a famous place, and people will ask to see your photos of it.

ANNA:

Pine Island is breaking up, has been for some time now, and when it's gone, that means perhaps even more warm water would reach Thwaites . . . I don't think things develop because of stability, I think the other way around. Crisis drives development.

Nearly a year later, Felipe and I cook an entire salmon in our new kitchen and invite friends over to share. We fill its stomach with scallions, rub its skin with garlic. Blacken both sides on our old propane grill. Between the main course and dessert, a fellow professor pulls out her phone and reads an email from the provost aloud: there is a single confirmed case of COVID-19 at our university. Campus will close a week early for spring break. Students will go home and not return. For a month or so, reports of the virus have appeared above the fold on the front page of the newspaper. First China, then Thailand, Italy, and Spain. It felt like something that happened elsewhere, until it started happening closer and closer to home. Washington, California, New York City. Now everyone seated around our dining room table in Providence inches their chairs back, increasing the distance between our bodies.

This will be the last time we see each other for a while. A week? A month? The future is suddenly very difficult to fathom, the path toward it even less clear. On the hillside, at the lip of that icy cave, I knew that to care for my child, I needed to go no further. It was a choice that, while unprecedented, came easily. This afternoon, this day, everyone goes home early, and I don't know what to do. I load the dishwasher and try not to touch the fork prongs with my fingers. Should I avoid the knife handles or the blades? To my left, my husband rinses plates and mixing bowls, dipping one after another, barehanded in a sink full of dirty water.

"You need to take better care," I snap. "Take my health and the health of your unborn child more seriously."

He looks at me as though I have lost my mind. He would never do anything to purposefully put us at risk. "I'm not going to be your punching bag, the outlet for your anxiety," Felipe barks back. It is uncharacteristic of him, but then so is everything. We finish cleaning in silence, sure that each of us is right in our own way.

The next afternoon I explore our new neighborhood, walk streets I didn't know existed. Down Bayard, which is shaped like an elbow, and up Creston Way, the incline causing me to pant. The few people I pass don't say anything at all, which just makes me feel more alone. My thoughts straying back to the medical evacuation, how quickly the initial shock sunk beneath the imperative to care for each other's well-being through simple acts like bridge lessons and lingering over canned lychees in the mess. The farther I walk, the more my thoughts mix. Just yesterday, I felt as if my belly button was going to turn inside out with the baby's kicks, and now the skin seems slack and the thuds slower, less frequent. Yet the idea of going inside a doctor's office fills my mouth with nervous saliva. How terribly difficult it must have been for Lindsey to be pregnant and sick and so far from reliable help. Suddenly I can't imagine it, even though I was there. Turn toward home, one hand fingering the discarded chocolate wrapper in my pocket, the other palming the dome of my abdomen, where my son floats in the dark. *Tell me how to protect him*, I plead with the slate-gray sky. *Tell me how to keep him safe.*

That night in bed, we take a kind of family portrait. Felipe's hand rests on my stomach. Our heads both outside the frame. My navy nightgown with white polka dots scrunched up beneath my breasts, my black-and-white striped underpants pulled down beneath the bump. It is the end of our first full day of isolation. There will be many more, though we don't know that yet. Don't know that our house has become a ship which we now steer alone, that the total number of days we spend there will eclipse the time it took to sail to Antarctica and back, that during this strange season everything and nothing will change.

That night, the first spring rain falls.

When I wake at two, I eat my buttered toast as an ambulance siren wails in the dark.

KELLY:

We're approaching the equinox. On the 21st of March, there will be twelve hours of light and twelve hours of dark, everywhere on the earth.

LARS:

Then it will only get darker, and darker.

KELLY:

I was out on the back deck washing down gear, and these shapes came at us out of the dark. You know, sea ice isn't as formless as you think. The floes and bergs are more like clouds. Then the ship would move away or change course, and the shapes would slip back into the night. It was super ghostly.

BECKY:

The ice has started saying, *You've got to go, now.*

A FEW DAYS LATER, RICK rakes the searchlight back and forth over the sea ice, hunting for leads. There is no clear way to move forward. It's nearly seven, but the world beyond the cone of light remains dark. He backs us up, stalls, then turns the thrusters on, and the ship plows forward again. The bow grinds up onto the floe, degree by degree, until the ice eventually gives way beneath us and a little crack opens. We squeeze into it, making the road as we go. But soon the floes pile up on both sides of the boat again, grabbing at the steel hull. Our pathway peters out.

"What's the coffee scoop doing over here?" Luke asks. He waves it in the air, and a few grounds fall onto the filing cabinet.

Rick sighs. "I guess I got a little discombobulated. But I think I'm finally getting you out of the soup."

"We're not there yet?" Luke says.

"We went up north and it was nasty. Over here we nearly ran aground." On the monitor, our tracks look like the wanderings of a lost penguin. "If we were on a sailboat, we'd be spending the winter."

Luke pulls out the Sentinel image and points to where Ferrero Bay opens into the Amundsen. "There's clear water up here."

"The ice is changing fast," the captain says, his voice bordering on agitation. After tagging seals, we spent a few days at Pine Island Glacier retrieving moorings, mapping the seafloor, and tossing the CTD rosette over the side. But yesterday afternoon, we decided to abandon the remaining science days and head north, for fear of getting stuck.

"How'd you sleep?" Luke asks me, changing topics.

"Me? I slept fine. You?"

"You must have taken an Ambien or something, 'cause Rick was crashing into the ice all night." Then he switches the Everly Brothers' "All I Have to Do Is Dream" to Joan Jett and the Blackhearts chanting "I Love Rock 'n' Roll." "I wouldn't want to see this in a week," he adds. The sky turns the color of reptilian stew, green and yellow with broad ribbons of burnished blue. A craggy old berg that looks like a middle finger rises off the starboard side. It takes nearly an hour to put it behind us.

When at last the sun comes, the sea is something else. As far as the eye can see, all is ice. Out on the back deck, I scoop up a bit of the snow that has fallen in the night. It occurs to me then that, after all this time, I still haven't really touched Thwaites. I lift my palm to my mouth and taste the flakes, but I know it doesn't count. I still haven't laid hands on the glacier that carried us here, and now that we are headed north, I never will. *How is it possible to have come so close and yet to have failed in this very basic way?* I wonder as I walk back inside.

GUI:

There is no culture here. Well, no human culture, other than what happened to the explorers. But that's not really a culture from here; it's an imported culture. The UK culture, the Norwegian culture, the US culture. When I see Palmer Land or Burke Island or Wilkes Land printed on a map, I think that these places should have a Chilean name or an Argentinian name instead. I mean, at least those are the two countries closest to Antarctica. Think about the name *leopard seal*. Who named the leopard seal a leopard? It was probably some explorer from the UK who went to Africa hunting leopards, and then he came to Antarctica and saw these seals and called them leopard seals. Animals have culture. I think the leopard seal should at least have a name all its own, not related to the travels of the human who happened to see it and to have the power to name it.

MEGHAN:

We're trying to find a way out.

ROB:

We have to go quite a ways west to get north and then quite a ways north through a gateway that's closing up to get to open water. It's time to get out of here.

KELLY:

I like the way places have been named after the first people who saw them because it's an international spread. There are Russians, French, Belgians, Brits, Norwegians. It's definitely multinational. There are places like Deception Island that have a descriptive name. I like those too; I like Shag Rock and Elephant Island. Then you get into the modern era, with the more recently discovered land features being named after a scientist, like the Bentley Subglacial Trench. You can piece together the human history of Antarctica from the place names.

JACK:

Wait, we're leaving right now? Seriously? I'm fucking excited. I was already excited about this being the last few days, but this is awesome. I'm about to bake the shit out of these cakes. I was already gonna do a good job, but now I'm like, *Hell yeah*. I'm going back there to tell Julian. He's about to shit his pants.

LARS:

This is how we humans work: we bring our baggage with us wherever we go; because there's no history here, you have to come up with something. I always talk about the Weddell seals in the Weddell Sea, and there you have a tongue twister. James Weddell—when he went into this area, he found these seals, in 1823 or so, and named them. The Weddell Sea got its name seventy or eighty years later. You might think it would be the other way around.

ALI:

In glaciology we have a whole world of words that describe the ice that don't apply to anything else on the planet. Like *grounding line*, like *ductile deformation*. It's quite a closed thing. There aren't more colloquial words to describe these phenomena. There's this whole history of what the ice has done that isn't actually written down anywhere, at least not in what we recognize as language. I interpret what I see in the sediment, in the seabed, from that I come up with a story in my head.

PETER:

I'm satisfied that we've had a productive, successful, dynamic, and exciting cruise, and now I'm also very ready to go have a glass of wine. There's a time for science and there's a time for not-science. Me, personally, I would not care to remember how many CTDs I did. Collectively we've done ninety-six so far. How many of those have my fingerprints on them? *Phffff*, I don't know, maybe half. A solid number. More importantly, though, my first drink is going to be a manhattan—or at this stage, let's be honest, basically anything with an ABV.

ALI:

My daughter is at a key age where she's very aware that I'm not home and is vocalizing it. Literally in the last two months, she's improving her language and saying that she notices that I'm not there. Sometimes scientists are painted as being not that caring, but we are all, literally, just to be here, sacrificing our normal lives for the sake of a cause we feel so strongly about. Surely that is the ultimate activism, to do the work that's going to provide the foundation for the answers.

KELLY:

Poor Ali, his daughter R— has gone from saying, *Where's Dad?* to, *Is he ever coming back?* to, you know, *Daddy is an elephant.*

RACHEL:

On the night shift, I was extruding one of the last Megacores, just siphoning off some water, when I looked down and saw a weird little gross globby thing in the tube. At first I thought it was an anemone, but then I realized it was a sea pig! It had a weird translucent body. Its insides were bright yellow. At the end of each of its legs, it had cute little suction cups that change colors with pressure.

We named the sea pig Dave. Dave, it was just what came to the top of my head. Then we realized it's International Women's Day and figured it should be Dava or something more androgynous. It was a sizable two to three inches long. Little creep. Imagine: it was just doing sea pig things, walking along, and then suddenly a tube happens to fall on it. I think its experience is similar to how we might imagine alien abduction.

GUI:

Remember the part in *Romeo and Juliet* where she says, *Why do you have to be a Montague?* And he says that he'll change his name; that a rose would still be red and smell good even if you call it something else. That's what really matters. These places and animals have their importance in their existence regardless of what we call them.

DURING THE FIRST CLASS I hold on Zoom, one of my students asks whether the virus has changed how I feel about bringing a child into the world. How easily the words flow out of me. "I can distance myself from the disease. There's money in the savings account and food in the pantry," I say, embarrassed. "So many people are so much more vulnerable." It is the same story I see playing out all across our climate-changed country, but at warp speed. Those who can afford to limit their exposure—by working from home or building a floodwall, for example—do. And those who cannot, suffer. This safety, for me and my unborn child, is the definition of privilege in the twenty-first century, though never before have I reaped its benefits in such a sudden, obvious way.

The days fly by in a surreal slipstream. Nights too.

In the absence of any reliable centralized response, we abandon the places that stitch our communities together. Schools, state parks, doctor's offices, museums, even borders close. The environmental activist group to which I belong temporarily disbands. A string is wrapped around the informal free library box three blocks from our house, with a sign in the window that reads "Lending suspended until further notice." Isolation becomes the primary means for many to express civic solidarity. Alone, those of us who can afford to be apart attempt to care for each other. Meanwhile, those who cannot remain isolated are forced to care in much more immediate ways for those who are ailing and for everyone else.

The sound of sirens increases; snowdrops give way to crocuses and daffodils.

When Chelsea, the midwife, finally calls me, I bombard her with questions about whether the hospital has run out of personal protective gear, how long I can expect to wait in triage before being admitted to a room, and if Felipe will be allowed to attend the birth. In the midst of her patiently explaining that things are constantly shifting, making it difficult to predict future conditions, the doorbell rings.

I startle. Then go back to listening. When it rings again, I waddle over to the front door and open it just a crack, fearful of in-person contact, the phone still pressed to my ear.

A man in his sixties holds up a cardboard box with an image of a car seat printed on the side. Behind him an Amazon Prime van sputters in the drizzle.

"I just don't want anyone to steal it," he says.

"Leave it there," I half bark, not wanting him to get any closer.

On the phone, Chelsea is saying there is a separate wing and separate doctors for COVID patients, that they even enter the hospital through a separate door.

When I turn back around, the driver is gone.

I leave the box—a virtual baby shower gift—on the porch so I can spray it down with disinfectant and give my complete attention back to Chelsea. Later, though, I keep looking out the window, wishing the delivery driver might miraculously return. I'm not sure what is more troubling: that my son's future safety has been purchased at the cost of someone else's now, or that there is no way for me to contact this thoughtful stranger, to thank him and to apologize for my sharpness.

GUI:

We're in a metal box in the middle of the ocean. You have the right to spend your days in your cabin reading a book and to come out only to do the things you need to do for work. But people don't do that.

Every time I need help, I ask. And every day, people help. Some part of it is boredom; they want something to do. But another part, I think, is the desire to contribute to something bigger than yourself. It's hard to say if life outside the boat could be like this, but we can learn from the experience. You can think back to the ship and say, *On the ship it wasn't like this; on the ship I didn't buy a bunch of random stuff to feel better; on the ship I helped other people; on the ship I was good at seeing what was important.*

JACK:

Turns out, being on a boat with a bunch of people I didn't know—well, it helped me grieve for my grandpa in a way. I had to work, had to keep going. And also I had you guys to talk to. Of course, I had my moments where I was just having an emotional time and crying. But remember when I did the BBQ pork loin? I know my grandpa *loved* that dish. And it was awesome to see you guys so into it. It reminded me of things he and I used to do together.

BARRY:

One of the nice things about being on the vessel is that it's one of the few places in the world where you have time. Time to do a job really, really well. Off the vessel, time is money, so you often rush through something just to get it done.

JOEE:

Have you ever used a well-made tool? It's a revelation. You practice writing—it's your craft—and your job is to make something beautiful and compelling, for the sake of aesthetics, but on the ship it can be a life-or-death thing whether someone is good at their craft. We live in a very encapsulated, handmade environment. Most wouldn't think of the *Palmer* this way, but human hands welded all these plates together. Human hands that knew what they were doing.

LUKE:

We use electronic and paper charts, rely on both. We also use iSailor, which is like a plotter. But by Thwaites, everything was off, because all of it was uncharted. So there we plotted our course by hand, penciling it in, recording our own depth contours.

JACK:

Can I tell you how much garlic I've peeled in the last two months? One day, I went through ten heads. I stood there and was like, *I have peeled so much fucking garlic.* And we still have four cases left. We're out of fresh fruits and veggies. You're the only one who knows this aside from us. We've got two bags of grapes. That's it. We're out of fish. We're out of whole chicken parts. We're out of honey. We were out of sugar, but Julian and I picked some up at Rothera. Thank God.

WITHOUT THE NEWNESS OF SHIP life and submarine test runs, without war room discussions around how best to use our time and informational videos about frostbite, everyone starts to go a little nuts. Some never take off their pajamas. Others sleep for fifteen hours a day. Bastien finally finishes *The Kraken Awakes.* Becky busts out her ukulele and I reread all of Ada Limón's *The Carrying* in aft control one afternoon. For three nights in a row, *The Life Aquatic with Steve Zissou* airs in the movie room, and though I hated the film when it first came out, now I feel in on the joke: a bunch of misanthropes attempt to make groundbreaking advances in marine science while simultaneously tending to each other and the machines that assist in their pursuits. Gizmos break, the boat's noises unsettle the sublime;

even their lab looks like ours, though their sauna comes with a masseuse. And on the *Belafonte*, it's the journalist who is pregnant, not the project coordinator. When the movie ends for the second time on the second night, someone hits play again.

Three days into our transit back across the Southern Ocean, I ask Fernando if he would like me to shovel the back deck. He's not allowed to approve such requests, so he refers me to Joee, who says no. But I'm too full of bottled energy to stand on the bridge wings and watch snow petrels scissor the air. Even the thrill of smashing through sea ice has dulled with time. Desperate, I tromp up the five flights of stairs to the bridge and inquire where my free snow-removal services might be put to good use.

"Can't think of anything at the present moment, with the science concluded and all," Rick says regretfully.

"How about the little deck off the crew break room? I passed it on my way up here. It's absolutely covered."

"Knock yourself out," the captain says, then he looks over at Rick, who backs off.

"Thanks!" I say. "I'll go ask Fernando to borrow a shovel."

After nearly two months onboard, I still haven't found many ways to directly assist the ship's crew. Surprisingly, there are more regulations preventing me from aiding those in the kitchen or engine room than in mud-core analysis. As I go back downstairs to get the shovel that Fernando couldn't give me earlier, I know this will be the closest I will ever come to helping him out.

"Here," he says when I find him, handing me the shiniest one of the bunch.

Thirty minutes later, my mind starts to relax into my body's rhythmic movements. I push the shovel into the top six inches of snow, then trudge over to the railing and lob the load over the side. Wait for the wet slap as it hits the water. Turn around and do it again. Soon I'm halfway down the bank piled by the door. *I'll clear the whole thing*, I think to myself. The sky is as wide as Louisiana, as wide as Montana, wider still. The farther away we sail from Antarctica, the closer I draw to my other desire. Becoming a mother is much nearer now than it was when we set out. The wind whistles around my ears, makes my exposed cheeks red. A big blue berg slides past. It looks like the clouds above it have cracked, a little lemon marmalade leaking through. Then go five Weddell seals on two adjacent floes. One flips on its back and claps its flippers together.

"Goodbye, seals," I say. "Goodbye, icebergs."

THE DEEPER INTO QUARANTINE WE sail, the more the separation between Providence and the *Palmer* collapses. The isolation, the boredom, the constant menace just beyond the window: all of it feels oddly familiar. I reupholster a brown glider a friend gifted us for the baby's room and learn to bake fruit tarts. Felipe roasts a leg of lamb, makes Bolognese from scratch. Each night at dinner, and in particular on Fridays, we talk about what meal we will make the next day, how to turn it into something distinct. I think back to Jack's unveiling of the king cake at midnight on Mardi Gras and the "carrot cake" he concocted for Bastien's birthday, appreciate them now, more than ever, as a way to maintain novelty when the weeks start sliding into one another.

Often I lie awake in the middle of the night, deep in worry. Sometimes I do not fall back asleep. Sometimes I hear an owl calling in the distance, hear the birdsong turn on as dawn draws near. Sometimes I feel my son hiccup inside of me, and I curl my body around his, a seed within a seed, the two of us sheltered for now. Our bed an ark, drifting through the pandemic's dark sea. When, at last, the sun rises, I watch the shadow limbs of the pines on the curtains, press my palm to my belly, and wait for him to respond, tracking how many minutes it takes to feel him move ten whole times.

"If you're going to stop coming to appointments," the midwife cautioned, "then you should at least do kick counts, so you might know if something's wrong."

Since the cruise, I've stayed in touch with many of my shipmates, including Lindsey. I planned to visit her during a recent trip to Oregon, but a snowstorm kept me from crossing the mountain range between us. In early April, I email her. Beyond my office window, the buds on the maple tree remain tight, reluctant to break. Though I don't say it in the message, the number of times I've felt the baby move is less than normal. Instead I ask after little Anvers, who was born in October, and tell Lindsey that when I feel unsettled I often think about her, how she navigated the Southern Ocean pregnant and uncertain and in pain. My anxiety must be only loosely veiled because her response is almost immediate. She writes: "I have no doubt you will get through this and it will be an incredible story to tell after the fact, just as my story gets told to people when they meet our perfect and happy little son." She tells me she is reading *Infidel*, a memoir by the activist Ayaan Hirsi Ali; that two of her moles fell off postpartum; and that Rob recently sent a note that said the scientists were able to complete most of their intended experiments, despite the medevac.

"I can't tell you how grateful I was to him for telling me that," she says over the phone when we chat later that afternoon. "To know that your need to care for me hadn't, in the end, undermined the mission's main objectives."

"I can only imagine," I say, nodding. We were lucky. Thwaites held her form just long enough for us to draw near. It feels even more miraculous now than it did then.

"Try to keep yourself distracted," Lindsey says before hanging up.

I take her advice and concoct a plan to build two raised garden beds. I call the lumberyard, pay for the wood, and send Felipe over to pick it up. A local farm delivers five cubic meters of fifty-fifty compost and topsoil, and I unearth a half-full box of deck screws from the basement. Every day after lunch, I take a break from work to continue the project. I predrill holes in the rich, red hardwood. With the jigsaw, I cut the spruce two-by-threes into eight-inch-long corner supports. Lean into each screw as I snug it into place. Then I methodically turn the earth, clearing it of weeds and roots and rock.

I plant the peas first.

At my husband's suggestion, as I drop each seed into its hole, I whisper a single word aloud. This is the word that will fuel the tendril's early growth. As the world goes to pieces around us, even faster than before, I find myself repeating the same phrase over and over again—seed by seed, hole by hole—so that it becomes a song for what will come. It starts with the nickname we have already bestowed upon our unborn son: *Nicolín, Nicolín, vamos a pasar el verano juntos.* It means, *Little Nicolás, Little Nicolás, we're going to spend the summer together.* At first, I thought the most important word in this incantation was my son's name. Then I thought it was the verb conjugated in the first-person plural. But the more seeds I press into their dark starting places, the more my focus shifts to the subject and the adverb. We are nothing if not in this together.

KELLY:

You know when you start hearing the same stories a second time that you've been together too long and people need some new experiences before you come together again.

PETER:

I want to see my cherished people. There you have it. I'm looking forward to stuff just getting back to normal, normal with all the goods and bads of normal: I miss bad days at the office, where you just sit there and pick your nose in front of an empty screen for two hours in the afternoon thinking, *What am I doing?* I miss that. It's been so long.

ROB:

Now that we're heading home, I can say I'm looking forward to getting some proper vegetarian food. There's not much cheese on this ship. I've usually got a stack of different cheeses in the fridge back home: blues, maybe some crumbly Cheshire, a soft Brie or Camembert.

SALAR:

I'm going to make a little origami fortune teller that will tell you which drink is going to be your first in Punta Arenas. Someone is going to get water.

JERMAINE:

Back home, my wife is coming along. We'll have a September baby. Her first trimester is almost over. In the beginning, she didn't like certain smells, didn't like certain foods, vomited a lot. But she's doing better now.

BECKY:

I just got off the phone with my son L—. He said he wanted to show me the two teeth he lost, again. He's been kind of obsessing over that. Then he said, *Mom, let me show you my skills*. He started doing all these tricks with the soccer ball. Before the internet cut out, he was dribbling backward.

JULIAN:

My little girl, she texts me almost every day. She keeps telling me she can't wait for me to come home. She's my little adventure partner.

JOEE:

One of the marine technicians on another cruise built a small collapsible boat while he was on the *Palmer*. When he got to port, he folded it up and shipped it home, and now he sails in it with his daughter.

CINDY:

I'll be going home to our farm in Salmon, Idaho. It'll be busy but not so different; I'll be keeping up with the house and cows and trucks and grandkids. My daughter just built a place down the street from us.

RACHEL:

> I'll return home a little different. I've tasted the Amundsen at this point. Electrolytes, vitamins. There are some minerals. It's really salty, actually, when you taste the sediment. It sounds obvious, but I've only tasted terrestrial deposits before, so this marine sediment—well, let's just say it's different. My mom must be so proud; her daughter grew up to eat dirt.

IN THE MIDDLE OF THE crossing, I walk past the whiteboard, where a strange drawing has materialized: a wizened old man with a long beard holding a trident. The caption reads "Neptune Is Watching." That night, I receive a message at my ship-based email address. The subject line: "Who Dares Incur My Wrath?" The sender, Neptune himself. It reads:

> I have observed that you have entered into my most Southern domain WITHOUT MY PERMISSION. My subjects tell me you have PILFERED my beloved rocks . . . STOLEN my precious water, DISTURBED my peace with loud tweeting noises, and GOSSIPED about your debauchery in YOUR PATHETIC HUMAN JOURNALS. Your awful conduct violates the time-honored moral code of the ocean and insults the memory of the truly intrepid explorers who have ventured south before you. You should have TURNED BACK long ago to avoid punishment . . .
>
> IF YE BE REPENTFUL AND WISH TO FACE THE ROYAL COURT FOR PROPER JUDGEMENT AND SALVATION then notify my loyal scribe, his Lordship Robert Larter the Terrible, of your desire to be saved. Once you have placed your mark,

you will receive further instruction pertaining to your official crossing ceremony.

HRM Neptunus Rex

I close the window and look around. The lab is nearly empty, save Becky, who is hunched by the bathymetry console with a Sharpie in one hand and a Styrofoam cup in the other.

"What's with the email from King Neptune?" I ask.

"You'll learn more in the days to come," she says cryptically, then returns to putting the final details on her meticulous line drawing of the *Palmer*'s back deck. In the absence of any kind of tchotchke shop, she has resorted to making souvenirs by hand. It is a bit of a tradition amongst marine scientists, who—instead of carving ships and lover's names into whalebone, as sailors might once have done—paint miniature boats and cruise numbers on polystyrene cups that are attached to the CTD console and dropped into the sea. During the descent to the bottom of the ocean, the pressure builds up, squeezing the air out of the Styrofoam. When her cup returns to the surface, it will be closer in size to a thimble. I walk over and see the words "Born To Core" wrapped around the orange cartoon sun she's painted on the bottom.

"Who are you going to give that to?" I ask.

"I don't know. I'll probably keep this one for myself. The rest are for my boys," she says.

I beg an extra cup off her and get to work making something for Felipe. I don't attempt to paint the *Palmer* or any of the activities onboard. Instead I write *2019 Desde La Soledad hasta Antarctica*—from our barrio in Bogotá to here—to mark the incredible distance we crossed in a single year; although my husband didn't join me physically, it still feels as though he has accompanied me on this journey. I decide to punch a tiny hole at the top, so I can turn it into an ornament for next year's Christmas tree. *Hopefully I'll be pregnant then*, I think as I draw squiggly lines around the jubilant lettering, King Neptune's message far from my mind.

Becky asks if I will be back for next year's cruise. She tells me that the THOR team plans to return to Thwaites and continue their coring program.

"As much as I'd like to keep following this story, I might not be invited," I say, pointing to my stomach.

IN THE BEGINNING OF THE pandemic, I jog the boulevard. When it becomes busy, I walk the grounds at Butler Hospital instead. The day a tent appears in the parking lot, I take one look at it, turn around, and never go back. For a while, I stroll the cemetery, the company of the dead more comforting than that of the living. But when the governor closes all public parks, it, too, becomes crowded, and so I start in on some of the least appealing streets in Hope Village, seeking solitude and air above all else. Every time I walk, I pass the shuttered salon and toy store, checking out my silhouette in the neglected windows' grimy reflection. Every day, the life within thickens.

AT LUNCH I LEARN FROM Joee that the original crossing ceremonies were for those sailing over the equator, from north to south, for the first time. The goal was to make the uninitiated feel as if they were going to die or drown—that way, the other sailors would know who was most likely to break down as the trade winds picked up. She tells me that men were tossed overboard, tarred and feathered, shaved bald, and beaten with wet ropes. Then Kiel, the third engineer, rolls up the sleeve of his blue coveralls to show me a black-and-white tattoo of a turtle. "I got this one after I crossed the equator," he says.

"Oh," I respond, unsure what to make of this information.

"You know, it shows that I'm not a pollywog anymore. I'm a shellback," he explains, as though that clarifies matters. "But don't worry. You'll be fine."

Tasha, who spent years in the Navy before becoming an oceanographer, leans in and adds, "You're absolutely not allowed to write about it. And no photos. What happens during the ceremony has to stay secret."

"But—" I begin to protest.

"You don't want to ruin it for everyone who will cross after you," she says. "The mystery is part of the fun."

"Fun?" I repeat. To which Tasha just raises her eyebrows and shrugs.

I consider arguing with her, insisting that this ceremony is important to the story I must tell, but doing so would just cause alarm. Better, I briefly think, to agree to maintain my silence now, a promise I can always retract later. But as soon as the treasonous thought occurs, I know unequivocally that I will do no such thing. Soon the community we have created will break apart. What better way to honor what we've become than by ceasing to act as its witness? After lunch I tuck my audio recorder into my desk drawer and thread the bungee cord back through the handle. There it will remain until we cross back over the Antarctic Circle.

Part Three | *Going to Pieces*

SETTING: The second time the author navigates the Drake, the sea practically holds still. "Drake Lake," the crew members call it. Up on the bridge, the captain and Rick talk about health insurance policies. Cindy prints detailed shipping labels for all the samples that have to clear customs and cross international borders to arrive at the right labs. Those with the British Antarctic Survey scramble to purchase tickets home as Brexit moves forward. The able-bodied sailors scrub the whole ship, inside and out, so the *Palmer* will shine in Punta.

The evening grows dark, but because we have crossed the convergence Luke doesn't turn on the spotlight. Icebergs aren't really an issue this far north. In every direction: liquid indigo turning black in the waning light. We stare at this immensity, tongue-tied and tired. Spent, we step outside to listen to the subtle shush of the waves and wonder if and when we will ever see water in all directions again; if and when we will ever again be held together in the ocean's glass eye.

"By tomorrow at this time, we'll be able to see the southern tip of Argentina," Luke says. "Me, I'm gonna put on my flip-flops and hit the Shuck Shack just as soon as the plane lands."

"I'm going hiking in Torres del Paine," says Meghan.

"I remember the first apple I had after my last trip to Antarctica." Joee seems to savor the memory. We four stand together, each of us lost in our visions of what is to come. For weeks, we've been talking about what we're looking forward to once we reach dry land—wine, solitude, sparkling water, drinking glasses as opposed to paper cups. But now that arrival is imminent, I think in the opposite direction: What will I miss once I get home? I climb the ladder into the Ice Tower where Becky plays her ukulele and recognize the opening bars from "Eleanor Rigby." When the moon rises above the clouds it shines a wide, wavering path all the way to the horizon. I look back behind us but there is no trail for my eyes to follow—no ice we parted—just the quicksilver sea, slightly rounded where it meets the sky.

"One of the cool things about being on the boat," Becky says, "is that when the moon's close to the horizon, it looks even bigger. I think that's Venus next to it, and over to your left, there's Orion."

I look out through the scratched plexiglass and piece together the three studs of Orion's belt. Search for Betelgeuse, the star that makes his shoulder shine, but can't find it. For a moment I forget that in the southern hemisphere, the constellations are all inverted.

"It looks like he's fighting the bear upside down," I finally say.

"The only other time I was here, I was quite young, twenty-three. It was before I had kids. I had just gotten married. It was like a dream to come here," Becky says. She stares past me, into the ocean. "I was so full of passion and excitement."

"I remember the first big mountain I climbed, around that age. I felt powerful afterward. I can't imagine what it was like coming to Antarctica."

"Right," Becky says. It's clear the comparison doesn't resonate. "It was also really difficult for me," she continues. "My stepmother had cancer and I was her caretaker, so me coming to Antarctica meant my sister had to take on that role. She was working a full-time job, while I was on this exotic adventure. That changed me. I've had a lot of experiences where I learned resilience myself, but facing resilience when it's for another person was really challenging. I found comfort in the ice and in the water. Watching the waves, feeling that kind of endlessness—that everywhere you look it's just the horizon."

We sit together for over an hour, talking about the cruise and motherhood mostly, about the amazing number of cores the THOR team secured and the strange sleep schedules that ground the first couple of months of a new baby's life. We talk carriers, breast pumps, and teaching schedules while the light of the full moon illumes our little room.

"When we started coring, I guess I got really intense, but we only get one chance and we need to get it right."

"One night we were about to do a Kasten, and you were sleeping, and someone said, *Should we really wake up Becky?* And Victoria was like, *Wake her up. She asked to be woken up. Wake her up.*"

"She knows I can handle it. I've dealt with three newborns. You don't need training to become a mother, you just do what you have to do. There's really no choice. The cruise isn't so different. Look at all we got done in the last couple of weeks," she says. "We sampled all the places we wanted, even with our very limited time. That was our team coming together, everyone on the boat, really. We knocked out a lot." Becky inhales slowly, drinking the air as though it were water.

My thoughts drift back to the stories of individual accomplishment that Antarctica has long been used to support. I think again how they encourage competition over collaboration, how they find whole groups to exclude. It is a narrative that protects the status quo, above all else, because it suggests that victories are achieved singlehandedly through genius and muscle and will, not through months and years of dogged work by the many.

"You know, dropping the Megacore is the most ashamed I've felt in my life," I say.

"Don't, I mean—"

"I'm over it, don't worry about it," I respond. This is only partially true, but I'm not bringing it up to be comforted. I'm after something else. "To have to face that and show up at work the next day, to own your actions and keep moving forward. It helped me understand what strength means here. Everyone makes mistakes, and the stakes are really high."

"I think we can move on because we have to be accountable. We don't get to *not* be accountable," she says.

"Wouldn't it be great if that were true off the boat as well?"

Becky's eyes catch mine and she nods. What neither of us says is that the world beyond the boat *isn't* the boat, no matter how much we might wish it were. That collaboration is easy when our numbers are small and there is no other choice. And that even here, even when we share a physical space for months at a time, still some work more in service of others. The engineers, the cooks, the sailors, the technicians—all attend to the needs of the scientists; our roles determined largely by the education and wealth we received off the boat, by the ways we have benefitted—even if indirectly—from extraction and for how long.

I look north, half expecting the dark tip of Tierra del Fuego to appear, but it doesn't. This is the last time, perhaps forever, that I will have such a view, so I try to savor the ocean's million obscure passages, how round and all-encompassing its reach. Suddenly one of Ali's seal-bark chuckles rises up through the stairwell, breaking my small reverie. Then Becky and I are laughing too. More than the bergs and the seals and the penguins and the sense of adventure that have punctuated every day of my life for the last two months, making them utterly extraordinary, it is my shipmates I will miss the most. The families that the bottom of the world gifted each of us.

BASTIEN:

You're less driven by your own individual desires on the ship, because, for instance, you can't go out and buy a new shirt. It's not like I'm craving a shopping experience. And while there are things you crave—a salad or a beer—there's just no way of going about getting them, so some part of that desire goes away too. In that way, life is easier on the ship.

MEGHAN:

I was picking Becky's mind a couple of nights ago about how to have family and be a scientist. I would never, in the office, sit down and quiz my advisor on the logistics of having children, but in Antarctica you're living with these people and barriers are broken down, like between professional and personal life. In the field, all relationships are more open. I think that's just what fieldwork does.

BECKY:

How do you adjust to the uncertainty that comes with fieldwork? Things go wrong and you have to live with it. It's similar to parenting in that way. You have to adapt. And you have to let go. That was me the other night. I got up to this point where I had to step back. I was so tired. I needed a few hours to play the ukulele and not talk to anybody. I had to trust my teammates. Trust in other people.

MEGHAN:

It's hard to imagine that we will run out of time on the boat, run out of time for these conversations, to work alongside each other all day, every day.

ALI:

I still haven't got sick of seeing anyone's face. We're all getting along as a family together, and that's actually fairly unique. We have so many shared memories that when we do get home, nobody else will ever be able to understand exactly what we saw and what we did.

JACK:

If I had lobster, you know I would be cooking it for y'all.

GUI:

I'm going to be missing Antarctica. When I think that next week, Monday next week, a week from today, I won't be seeing any of you or the *Palmer* or the ice for at least a year—well, it's sad. And it's crazy, because I've never been here before. It's all new and I'm already missing it. It's like falling in love. When you leave, you want to see that person again, like right away, even though you just left.

It's so rare that we drive anywhere these days. When I do get in the car, I often discover the seat tilted way back, a sign my husband has been using the space for his weekly therapy sessions. We conserve the most basic things—squares of toilet paper, conditioner, tea—attempting to stretch out the time between grocery runs. There is a brief foray into jigsaw puzzles, a phase where we wear our winter gloves everywhere that isn't our home, and the persistent feeling that plans constantly shift. Time changes shape again, pulls in even tighter around my transforming body. My breasts grow heavy, folding into my chest, and the baby turns head down, preparing to pass through the birth canal.

We all do what we can to try to hold things together. We bake brownies and drop off half on a dear friend's porch. Sneak a sachet of Siberian wildflower seeds into the mailbox of another. Create garden kits for those who live nearby: three different tomato starts and a clump of mint and basil sprouts. Tend to the peas, kale, and lettuce popping up in the new garden beds, unsure whether we will need the food come summer. A friend bakes an extra loaf of parmesan bread every week, which he delivers. My yoga teacher drops off a bag of hand-me-downs for Nicolás. We receive children's books by mail, and hold weekly Zoom dates with family. Every night at seven, the local fleet of fire trucks parade down Hope Street in tribute to all of the first responders and health care workers who continue to care for us by putting themselves at risk.

Sometimes, my stomach turns into a tortoise shell, my uterus practicing contracting. I eat prunes and soak in the tub at night. The bathroom window is nestled in the crown of the two-hundred-year-old beech, the reason, we half joke, we bought the house in the first place. We watch in quiet amazement as its buds burst open over the course of several weeks. It becomes one of the many small rituals that ground our days, like how I used to start almost every morning on the boat up in the bridge drinking tea, watching the ice, trying to note its hourly evolution. In this way, and so many others, it's almost like being on the *Palmer*.

One rainy evening Felipe and I construct the crib where Nicolás will sleep. As we work, we talk about the place where I will give birth—a small country hospital on the state's northern border, a decision we made to reduce possible exposure to COVID—and about my new ob-gyn, who cracked enough jokes during our first meeting to leave my cheeks sore. We talk about how the mayor has turned certain streets into pedestrian paths on the weekends, and about the dropping carbon emissions after the sudden halt of so much economic activity; we talk about how, to keep them low, we need any potential stimulus packages to set aside money for the transition to green infrastructure, and about how, early in the pandemic, the organization where I volunteer pressured the governor to enact a moratorium on utilities shutoffs—which she did, and then thanks to public pressure, extended. We talk about all the possibilities for change set in motion by this sudden upending event, and about how quickly it has further exposed the deep structural inequality that shapes our present.

"What any of this means is up to us," I offer. The sentence sounds good when I say it, but I also know I have failed to clarify the most important word. Is my "us," to use the lingo of the Occupy movement, the 99 percent, or is it those who hold positions of power in our country at the end of five long, damaging centuries? If it is the former, is our failing democracy even able to honor and respond to the majority?

Outside the sirens wail, while inside Felipe hands me a screwdriver.

"I'm not sure if that's reassuring or terrifying," he admits.

"Both," I say. "It's both."

I look out the window and long for the boat, long to labor alongside other people. People who were born in so many different places and times, people with different hobbies and even wildly different political leanings. There was something satisfying and hopeful about it all.

MARK:

I don't think my birth was particularly notable. It sounds a bit sad, doesn't it? I can guess it would have been reasonably warm, because I was born in Hong Kong in November. It was definitely sunny. I was the first, and there's one more like me. I have seen photos of having just gone back to our flat. There are some of me and one of my mom without me, which is pretty cool.

ROB:

What do I know of my own birth? It was the middle of the day. That's all I know. I think it was one o'clock in the afternoon. My mother was an older mother. I was the first—the only—child she had, at the age of forty-three. My partner and I, we have three. She was thirty-five when she had her first, forty when she had the last. She's a young-looking person for her age. Unlike me. I've aged and she hasn't.

I was born in Wolverhampton, West Midlands. You can't get more central in the country. It was one hundred miles to the sea in any direction. Which doesn't sound like much if you're from the middle of North America. The thing was, people didn't go to the seaside where I grew up. You went for your once-a-year holiday, maybe, to the seaside. We didn't have motorways in that day—or what do you call 'em, freeways? We didn't have six-lane motorways. Any journey involved going through town centers on a road that was one lane of traffic going either way.

JERMAINE:

I didn't ask what my birth was like for my mom, but the one thing she told me is that when she was pregnant, she really liked coconut juice. She craved for coconut juice, always. And now I like coconut. I was born May 13th, 1985. In the Philippines, May is a beautiful month with many flowers blooming. I'm a Taurus, strong. I do feel like that, that I am strong.

JOHAN:

I was born in the geographical middle of Sweden. It's called Östersund. It's the same latitude as Anchorage, so it's a bit up there, but it's not the Arctic. I don't know much about my birth. I know that my mother was a very tiny woman and I was a very big baby. I was four-point-something kilos, which apparently is a lot.

ANNA:

I should ask my mother. I've not been told so many stories. I was born in a hospital, everything went fine as far as I know. It did not take very long, and there was not too much pain that my mother remembers. No horror stories or funny things. Just a regular birth, it sounds like to me.

Talking about birth stories, I think it's interesting to think about how, when ice recedes, we get more land down here. It might cause sea level rise up north, but down here the earth rebounds. Something they've started to talk about is the greening of Antarctica. Antarctica has started to grow things on the peninsula. When I looked at the Lindsey Islands and Edwards Islands and Schaefer Islands, they were totally barren. There was lots of energy stored in the penguin poo that covered them, but no plants or people have used it yet. It's energy just lying there. For example, in one hundred years, maybe there will be things growing on those islands. More life, maybe new species.

Our last evening on the *Palmer* feels like a party, or as close to one as you can get when the lights are all fluorescent, the drinks nonalcoholic, and the galley the only thing vaguely approximating a watering hole. By this time tomorrow we will be docked in Punta Arenas, and the guarantee of each other's company will be gone. A handful of people cluster around the toaster oven and debate the merits of Marmite. When I admit I've never tried it, Ali butters a slice of wheat bread, slathers on a layer of the yeasty concoction, and hands it over eagerly. Though I want to like it, I cringe at the earthy, engine oil–like spread.

"Maybe it needs a little bit of honey?" I say. But then I remember that we've run out.

"You know, Scottish bees, they've never been bred to bring out domesticity, not like English bees," Lars says, his shock of gray hair bouncing. "Sometimes when I work with them, I imagine them donning kilts and jumping out of the hives, battle-axes in hand."

"You're also a beekeeper?" I ask, incredulous. Only recently did I discover that Lars throws pots, a hobby that replaced his passion for speed metal back in the nineties. He tells us that native bees evolved to cope with the islands' wet winters and unsettled summers. Ali finishes my slice of Marmite on wheat and Joee nibbles a fistful of dried apricots.

"Remember Marjory from *Fraggle Rock*?"

Of course Joee watched Fraggle Rock, I think. It's been well over two decades since I last saw the show. "The singing trash pile with the sunglasses?" I half-ask, conjuring up a vague memory from childhood.

"Yeah. Like, who decided that the town's garbage would be a character in the first place? She was way ahead of her time." We talk about the genius of Jim Henson and try out each of the remaining jams in the larder. My vote for the best goes to the lemon curd, which I plop on another scoop of frostbitten vanilla ice cream. At some point I notice that we've broken the mess's cardinal rule: together we occupy the far end of the table by the wall, usually reserved for the captain and crew.

"We have this really incredible connection on the ship, right? We even share the same dreams and aspirations. But if we meet in two weeks' time, on land, some people will be different," Lars says. "Which is really scary. I hate that about the end of a cruise."

Beyond the porthole, the inky outline of Argentina slips past in the night.

Ali reaches across the table to grab the squat jar of Marmite. On his ring finger he wears a band of light rose gold that looks more like a piece of hardware than a symbol of matrimonial commitment. "I've never met anyone whose wedding band looks like mine," I say, holding my hand alongside his. The edges of each ring are not rounded but flat, giving them a practical, utilitarian appearance. "Becky's got a similar ring too," I add. "There are three of us on the ship."

We talk about how the continent attracts people who are considered a little odd back home.

"I do feel like there's a relationship between those who are sort of relegated to the corner on the mainland and a knack or need or ability for the creative problem solving that Antarctica demands," Joee said once during transit. Now she shoves her hands into her Carhartts and adds, "There's something about life on the ship that is its own kind of craft making—like, we construct our community with a lot of care because it matters to each of us, for survival but also because we need each other in this other way too."

"Anyone want to play cribbage?" Scott asks. "If you don't know how, I can teach you. It is"—he pauses, looking around to make sure Anna isn't in earshot—"way easier than bridge."

Some folks join him, others stay behind. I make popcorn in the microwave oven and dip the hot kernels into the slab of Nutella I've spread on my plastic plate, happy to bask in the easy company of these people who, just two months prior, were complete strangers. Though no one says it, we are all, in our way, attempting to savor the sweetness of friendships forged in the crucible of ice. While I'm not sure when I will see anyone again, or how changed we will be if we do meet, I do know that back home I will begin the process of writing a book about the journey we took, and that its pages will keep some part of our community connected long after landfall breaks us apart.

THE AUTHOR:

On May 26th at about one thirty in the morning I woke up. But I thought, *This can't possibly be labor*, because you were two weeks early and it was my birthday. I just couldn't fathom that you would come on my birthday. Then around 5 a.m. I took a shit, and when I looked down I had also pushed out this kind of translucent white slug. I sent the doulas in our group chat a photo, and they were like, *Yep, that's the mucus plug*. That's when I knew I was in labor. I thought, *Okay, it's happening*.

I thought about all the things we didn't have in the house. 'Cause it's—well, we're in the middle of a pandemic. We only leave the house to go grocery shopping about once every two or three weeks, to limit our contact with other people. We were at the end of our last supply. So we didn't have anything, really, in the fridge. I wrote the doulas again and said, you know, *Does Felipe have time to go grocery shopping?* I told them the contractions were about twenty or thirty minutes apart. They said, *Yeah, he can go, this part can last a while*.

He left and I stayed in bed. Mostly. I also sprayed witch hazel on a bunch of maxi pads and put them in the freezer. I don't really know what happened next with the groceries. I mean, Felo came home and we unloaded everything that he had gotten: pounds of cut-up watermelon and a big juicy steak for when we returned from the hospital. We asked the doulas what to eat and they sent us some recommendations. I settled on scrambled eggs, a piece of toast, and an avocado. I ate that, and it made me feel strong again.

We moved our little unit outside under the beech tree. Between contractions I would sit on the yoga ball and talk with your Papá; he would tell me stories to keep me distracted. There was even a period when I finally opened my birthday cards. We hadn't gotten to that in the morning, in the rush to go to the store. So I would open one card and then have a contraction and get down on all fours on the yoga mat. As soon as it stopped, I was fine. Which kind of amazed Felo. He was like, *It's over?* And I was like, *Yeah, I guess it's over.* The pain would disappear so fast. And then I would open another birthday card.

Sometimes I would kick out my left leg. That was my instinct, to kick out my left leg. When I told the doulas that I was feeling a little stuck, they said, *Okay, between two contractions, why don't you put your left foot up on something a little bit higher and lunge into it?* I walked over to the beech tree and I put my foot up on its roots. I leaned into it and I put my hand on its beautiful elephant-skin bark. When I looked up into the canopy, it reminded me—I don't know how to put it—calm ran through the tree into the palm of my hand, into my body. I knew then that I wasn't the first person to labor under that tree.

At one point, the birthday cake came out and I had a big piece. It was an ice cream cake. "For Mamá from Nicolín" was written on the top. That was my first gift from you. And I thought to myself, *I may be seeing this again, but that's okay.* It tasted delicious. Full, full of love, full of love.

I never felt scared. I never felt overwhelmed, I felt very present. Like I could do it. I never doubted that for anything. And I knew, even then, I couldn't do it without the strength and support that I draw from all the people in the world who have loved me beyond the beyond, since the very beginning. That's what Felipe and I want to give you—that feeling of support and love. 'Cause I can tell you, when you have it, you grow so strong inside it.

It starts to get dark. I think it must've been, like, seven, and I got cold. So I said, *Let's move onto the porch.* I honestly don't know how long we were there. At one point, Felipe put a hot compress on my lower back and an insane cascade of water poured out of my body. I remember being so disappointed that I had just put on a clean, dry pair of pants and now they were soaking wet. It was *so* much water. I could hear it dripping through the floorboards onto the ground beneath the porch.

After that, the noise I made wasn't a moan anymore. It felt like I started riding the sound, the growl emanating up from my gut, that I had to—every time the contraction came, I had to find that sound and meld my voice to it. My voice would warble until I found it. And then I would hold onto it for as long as I could.

What happened next felt like a crisis. But Felo was still very calm. I looked at him, and I was like, *We* need *to go to the hospital, and I* need *a sweater and a nightgown 'cause what I'm wearing is soaking wet.* He gathered the last couple of things that we needed, and Elise and Christopher, who were driving us to the hospital, showed up.

I remember walking down the stairs, and as I got outside into the night air—it was night now—I could hear the neighbor on his porch playing the harmonica. It sounded like he was playing for us. I mean, he could obviously hear us. There was no way you couldn't. And he sat there, in the darkness, playing. Then a contraction struck and because my instinct all day had been to go on all fours, I doubled over in pain. I was in a flowy dress at this point and my bare ass was wagging around in the air in front of our house. I sensed someone—you know, a stranger—walk past, and I thought, *Well, that's something they'll never forget.*

We took off through the night. Christopher drove with incredible urgency. I think he was afraid that we were going to have you in the car. He was wearing an N95 mask. But when I looked at his face in the rearview mirror, I could see that his eyes were wide open.

I think I had like five or six contractions in the car. Maybe more. During them, liquid poured out of my body. I didn't know what it was. My vagina was just sliding around against the leather. I had no idea what was going on down there, but it felt like a mess. We got very close to the hospital and a yellow light turned red. Chris stopped. I really wanted him to run it, but instead of saying that, I made a joke. I was like, *Well, the back seat of this car is never going to look the same again.* And Felo said something really beautiful about it being the place where we started bringing the family, you, into the world. You know, your Papá is like that, a very profound guy. I remember saying, *No, I mean, the seat is literally covered in urine and blood.* I don't know if anyone laughed.

Finally, we got to the hospital and the front door was closed. I rapped my knuckle on the glass and then I dropped to all fours right there in the vestibule. They got me into a wheelchair and that felt horrible, but also I have no idea how I would have walked. We went in an elevator up. They wheeled me into a room, and I don't know how I got on the bed. They were checking—between contractions—my blood pressure, my temperature. At one point I had to have a COVID test. They said, you know, *We need you to sit up very straight.* I was on all fours this whole time, and I had to come up onto my knees, and they were like, *Now we need you to hold still while we shove this thing deep into your nose and you're going to cough a little bit.* I was like, *Fine, whatever.* That was the least of my concerns.

The pain was riding me now through new sounds I had never, ever made before. Instead of riding a wave of sound to survive, um, I had to somehow summon sound from within and then strengthen it with energy from my body. Grab that sound from deep, deep inside, and then push it out with as much force as I could possibly muster. And usually by the end of the contraction, I had run out of energy, and the sound, the noise—no—the roar would go up in pitch. And then I was sort of yelping like a wounded animal. I felt like I wasn't moving you along then. I was just in so much pain that I couldn't do anything but yip and wail.

I asked, *When do I push?* And Dr. Morton said, *Not quite yet.*

They got a big plastic pill in front of me. And I would lay my head to the side and just, like, rest my head on it between the contractions and breathe. Amazingly, Lisa, the doula, showed up. Everyone was wearing masks and she was beside me. She and Felipe would take turns, you know, putting a bottle of water in front of me with a straw so I could drink.

I remember someone telling me that if I wanted to have a voice the next day, I should scream a different way. And I thought, *I really don't care if I have a voice tomorrow.* What else? Rachel, the nurse who was attending the birth, was keeping a warm compress on my vagina to help it stretch out. Lisa put a cold compress on my neck. And when I would have a contraction, she would put another one in front of me and I would bury my face in it.

Then they took the pill away, and I was as low to the bed as I could be, with my ass in the air and my face buried in the mattress. Inside the contraction and outside the contraction. Rachel kept telling me to move my energy, move the growls, move everything into my butt. Dr. Morton said, *Pretend like you're taking a shit.* Then he paused and said, *Like, take a shit.* Someone needed to say that to me, 'cause I think I was still nervous about having my butt in the air and the possibility of pooping in front of everybody. But when Dr. Morton said that, I felt like it gave me permission to actually take a shit. I had no idea what would happen if I pushed in that way.

It's funny. Everyone tells you birth has all these different stages, like there's early labor and there's active labor and transition. And then pushing. And I remember kind of wanting to know what stage I was in. Was I in transition or pushing? But no one was using any of those words. So I was just in what I was in. I couldn't be anywhere else. It's like you really don't know anything about what's happening, but you also know everything about what's happening.

I felt my uterus clamping. The contraction would hit. And then this other muscle in my stomach would squeeze super hard and I had to squeeze with it. I did that probably three or four times, and then it started burning. I remember thinking, *You've got to be kidding me—on top of everything else, now my vagina's burning?* But it really only happened maybe two more times, and then it was over. Lisa kept saying to me, *Bring your baby down, keep bringing your baby down. Bring your baby down.* I was struck by that clamping urge, and I pushed down with my muscles that I use to poo and thrust my butt back and just tried to make the most of it.

I remember thinking, *I don't know how much longer I can do this*. I was getting exhausted. And then I thought, *Okay, if you want him to come, you have to push with all your might. You can't save that energy for later, just because you think you might need it. No, throw everything you've got into it right now*. And I pushed—I don't know—three separate times after that. The second time, I felt your head come out, and the third time, I felt your body come out, and then you were with us.

I heard someone tell me, *Bring him up to you*. So I reached down between my legs. I carried you under my body, then many hands were on us to help us turn over. You just put your head right there on top of me. And it was absolutely awesome. Awesome. That's the word. Truly just awesome to have you in my arms.

THE LAST MORNING ON THE ship, just before dawn, I undog the door and step into the velvet dark. The air is dense with the musky smell of earth. For months, the little scent the wind carried was hard-edged and blue. Now this. I inhale deeply the smell of decomposition, the more it makes. The chalky outline of Patagonia, those round rocky outcroppings, tumble toward the sea, turning pine covered and disarmingly green the closer they get to the water's edge. Just off the back deck, a giant petrel curls close to the ship, then peels away again, out into the dawn-ripe air. His flight parabolic, like the symbol for infinity. He flaps his wings, once, twice, then sets his figure into a taut *W* to cruise. His body a tool that has evolved over millennia to let the wind hold it aloft.

The door to the deck is open, so I can hear when a call comes in over the radio, asking for our anticipated time of arrival in Punta.

"Seventeen hundred," Rick says. His disembodied voice crackles against the sky's evolving mix of periwinkle and pink. Where there was one bird, now there are two. Both with wingspans almost as wide as my body is long. I walk up to the top deck to get a better view and discover Gui there, using the smokestack as a windbreak. Together we watch the birds draw figure eights behind the boat, their flight weaving the sky and sea into an elegant conversation. Then one bird suddenly pulls away, winging toward what we cannot know. His body momentarily obscures part of an offshore oil rig, which from this distance looks like a child's backyard fort. It's the third one we've passed this morning.

"For months, our smoke has been the only thing in the air besides the birds," I say. Now I can't tell whether the faraway plumes of gray on the horizon belong to us or someone else. But their exact source doesn't really matter; it doesn't matter if they billowed up from the *Palmer* or from the nearby Gregorio refinery. They are ours no matter what—whether we sail to Antarctica or stay home, whether we have children or tend to those of our friends—no individual choice will keep our human imprint from the air or freeze Thwaites in time.

The second petrel pulls back into my plane of vision. Gui takes out his camera and declares he has to make a video, even though he knows it won't do justice to what we witness. He leans against the steel to steady his arms, then begins to follow the bird's graceful flight. What is happening is both very new and very old: the petrels whir westward through the Strait of Magellan, carried in the wind wake our ship makes.

"If I weren't working with marine mammals, I'd probably be working with birds," Gui says as the giant petrel's shadow passes over him. "I think it has a lot to do with my imagination. I like to imagine the intelligence of these birds, their ability to cope and adapt to different environmental pressures, to cross the Drake and breed in Antarctica. It's pretty amazing that they're here, surviving and reproducing."

I stand next to Gui, awestruck by the petrels and the petrichor rising up from the sun-warmed land. Everything and everyone has a beginning: a being that begot them, a place from which they start. And yet the care that carries them into and through the world isn't the work of just one pair of hands or heart. I test out different words that might describe not only birth, but how dependent that process is on a whole web of relations between sun and rock and sky, between all of us. *Procreate*, *propagate*, *breed*. Each sounds too narrow on my tongue, too tied to one body producing another. That's when I arrive at a word that has been with this project from the start: *regenerate*. I know that my life is a destructive force, but I am also trying to tend to the people and places that sustain me, to carry some of that old magic forward. Because if I wish a child into this world, then I must also wish this world upon them.

THREE HOURS LATER, WE ARE back in port. I eat one last meal onboard. Already the changeover has begun. People I don't know ask about the mission while I chew through a plate of fish and chips. What can I tell them? Nothing hits the mark. Luke and I have three beers at the Colonial. I'm not ready for it to be over, not just yet. Then I join the scientists at a local restaurant where we eat *merluza* and drink wine. Rob wears a button-up shirt saved just for this occasion. Some people are already gone, catching the first flight out of town. Afterward Anna and I close the night at a tiny bar. We smoke cigarettes under the streetlights and lean against each other, triumphant. Some sleep on the ship one last time. I set my alarm, catch a taxi to the airport, and fly to Santiago, where, thanks to the previous night's drinks, I throw up in a trash can. At first, I am afraid someone will see me, but every single person who streams past is a stranger. That's when I know that the very specific thing we made has ended. That it won't come again.

Epilogue

SETTING: Almost a year to the day after setting sail (strange how this keeps happening, how one January mirrors another), the author is in Oregon, nestled into the side of the Cascade Range, her cabin at the limit where rain turns to snow. There she works on the first draft of this book, a book that will tell a story about motherhood and glacial retreat. A book about a community that is falling apart even as it comes together. What precious days, each mist-shrouded, jade benediction, tiny gems to be savored before her time becomes less her own. At the end of the residency she and her husband drive down the mountainside, snaking along the McKenzie River, toward the library of mud, to visit the cores that were gathered and sampled during the cruise.

The world's largest collection of geologic material from the Southern Ocean is housed in a low-slung, rose-plated building nestled among an urgent care clinic, a Safeway, and the local branch of Habitat for Humanity in Corvallis, Oregon. Val Stanley, the Antarctic core curator, greets me in the foyer, her lightning-bolt earrings reflecting the flickering fluorescents overhead. We spend a few minutes making small talk while Rodeo and Bella, the two lab dogs, lean their furry bodies against Felipe's legs. As someone who often works in archives, he is excited to experience a wholly different kind of collection, one where the documents are marine sediment samples retrieved from the bottom of the earth, not the fragile folios of nineteenth-century periodicals he studies.

"If the mud is the text," he asks, "then who is the author?"

The question takes me by surprise.

"Thwaites," I finally say, "and, increasingly, us."

Val pulls her tangerine scarf tighter around her neck. "When Oregon State started to collect sediment back in the sixties, the samples were stored in the walk-in refrigerator of a famous Chinese restaurant downtown. We've since expanded," she says. "Want to take a look?"

"Of course," I respond. "But let me run to the bathroom first."

Afterward she leads us through a room that looks a lot like the Dry Lab, with two wide white tables running down its center. Then we enter a bigger, more cavernous space with a gigantic sliding metal door on one end. "You could easily park a 737 in here," Val says, pulling it open. "This used to be a factory where they built ink-jet printers, then it was a plastic mold injection plant. The university bought the place just after the 2008 market crash, in a kind of opportunistic moment. Today we store the largest collection of piston cores in the world, among other things."

Together we enter a cold chamber that reminds me of IKEA. The floor and ceiling are a light granite gray, the long rows of metal shelving bright orange and blue. The air temperature hovers just above freezing to keep the critters suspended in the mud from decomposing. "The oldest Antarctic core we store was pulled in 1962," Val says. "Think about just how much our understanding of the continent has evolved since. Back then they were trying to answer a really basic question: What is Antarctica made of? And now we're trying to piece together how, for instance, the West Antarctic Ice Sheet responded to past spikes in CO_2."

"Are there cores that lots of people come to sample or see?" I ask as we walk deeper into the room.

"Celebrity cores? Yeah, we've got a couple. Take ANDRILL here, for example. It's the longest in the collection, containing a slice of earth history that reaches back seventeen million years," she says, gesturing to an aisle lined with what looks like hundreds of pizza boxes, each stacked four deep, yard after yard. Val tells us that it took scientists from four different countries two field seasons to coax this stone from the earth, creating one of the most complete pictures we have of the Pliocene (three to five million years ago) and the Miocene (fourteen to fifteen million years ago), two periods when the planet was significantly warmer than it is today.

"Something I'm always thinking about when I do archival work," Felipe says, "is what's been left out. When you work in an archive, you're always only seeing what the author and the curator wish, only hearing from the voices they include." His words make me think back to my initial library trip two years prior, the one where I discovered that most of what has been written about Antarctica was penned by lone, enraptured men. How small the collection was, and how boring, for they left out so very much.

I turn a circle in the library of mud, taking in the thousands of plastic tubes and boxes of sediment samples. Back when I set out, I had no idea that this other archive existed. It, too, is curated, but its contents are dictated by the questions humans have chosen to ask about Antarctica, questions that are shaped as much by the pressing concerns of the day as they are by the culture of the geophysical sciences, which for most of the last two centuries was also a man's domain. But that is also changing: the improved gender parity of our cruise a sign of this shift.

"At least here the voices rise up from the earth," I say.

"I can pull sections from ANDRILL if you'd like," Val offers. I realize then that she doesn't know what to do with me since I am not, unlike the other visitors who wash up on her poured-cement shores, here to sample core.

"If the tubes are arranged chronologically by sample date, that means the most recent *Palmer* cores should be closest to the door?"

"You're a natural," she says.

"Let's go look at those."

We turn around and head back in the direction from which we came.

Tap, tap, tap. From deep within me comes that otherworldly Morse code. I cup my hand over my belly while Nicolás pushes his feet against the walls of my uterus. Val opens her mouth to speak, then stops.

"I'm pregnant," I reassure her.

"Oh, congratulations! I suspected but didn't know if I should say anything."

"Secretly, I came all this way to touch some of that Thwaites mud with this little widget inside me," I offer. All week long I have been thinking about how doing so would allow me to close one circle and open another. There are so many ways in which my journey toward Thwaites taught me how to mother—or, at least, how to invest a whole lot of time and energy into a project without having any idea how things will turn out.

"Here we are," Val says.

The twenty or so sheaths pulled from the ocean floor during our time on the *Palmer* seem fairly insignificant when compared to the sheer volume of mud and rock surrounding them. And yet I know that each contains information that exists nowhere else on the planet, is itself a unique record of the many ways Thwaites fell apart in the past, which means it is also capable of helping us understand what is happening now. I ascend the industrial stepladder until I stand face-to-face with some of the very samples I helped extrude on the *Palmer*. Slowly I run my hand over the cores. When I open my mouth to say something, no words come out.

"I pulled the cores you requested," Val says below, indicating a rolling dolly where half a dozen tubes have been set aside. I descend the five steps to the floor and follow her to the core-cutting room, where she lifts KC04 onto the table and slides it from its new protective case. She removes the plexiglass lid. The mud beneath appears almost as damp as it did on that first morning when we hauled it up from the bottom of the Amundsen Sea.

"The sediment looks warmer, somehow."

"It's oxidized a bit. When you got it onboard, it had never been exposed to air before—now it has."

"Has anyone from the cruise been by to sample yet?" I ask, looking over the core again. It appears as though some mud has already been removed.

"Oh yeah, a whole crew of people came through in October," Val says.

"Like who?"

Some of the names I don't recognize—they weren't on the *Palmer*—but others I do, like Rachel, Victoria, and Becky. As soon as she says their names I miss them deeply.

Down in the dark of my womb, the body within my body turns. The sensation is strong—like thrum, like heart clench. I have often thought, in the months since leaving the ship behind, about the radical accountability that defined that period in my life. A thin piece of Saran Wrap separates the sediment from the air.

I lightly run a finger down the core's length, leaving the faintest imprint behind. We lived alongside one another; we had no choice. Proximity made such care possible. Now my son is the one teaching me that there is no end to what we give one another and to what we owe, that every single being on earth was born through the innate generosity of a body. A body as large as the last continent on earth, even larger than all that. We have a responsibility to one another, beyond blood ties and genes, to make good on the gift that was so freely given. This responsibility—to the matter out of which we arose and to which we will return—remains constant.

"I sort of feel like I'm being reunited with old friends," I finally say, and by this I mean both with the mud and with my memories of my shipmates. I can tell from the look on Val's face that she understands. Then I pick up one of the two Megacores from the same location as the sample I dropped.

"There's no other . . ." I pause, not knowing how to ask after the mud that trickled through the floorboards in the Baltic Room. "There are no other samples from this location?" I punt.

"There's a Kasten," Val says.

The archive notes no missing cores, which means there is no mark of what I lost. My mistake has gone unrecorded, in part because what we were able to bring back was enough. I clasp my hands around the sheath, letting the coldness of the mud work its way into my palms. This is but a tiny fraction of what the seafloor in front of the world's widest glacier contains, and yet it is also an entry in an archive written in a language that comes before our own: one that is vast, produced by collective effort, and open to anyone. It is an archive that reorients our attention, focusing it on the story Antarctica is telling us rather than the one we want to tell about it. Our task is to draw as close as we can—in spirit if not in body—and listen. To act as though our lives were unfolding alongside the ice, always in conversation with it and all that it contains. I am less sure now than when I set out that we know just how to do that. Though we should try. We should definitely try.

When I began this book—a task that started when I wrote a proposal to be sent to Antarctica as an artist working in service of the science conducted there, a task I thought I would tackle by talking with the people who know the place best—I assumed, incorrectly, that these scientists' stories would communicate the radical transformation unfolding there, largely beyond most people's sight. I thought that I, too, could go and come back and carry a little of this knowledge in me. That I would witness change and that it, in turn, would change how I went about living my life. But when a single person looks to Antarctica hunting for signs of transformation, what that individual sees is but a tiny fraction of a much, much larger picture.

Imagine that you are looking at a northern forest instead of a glacier's face. Your gaze might fall on the soft bough of a hemlock tree, its underside spotty with egg sacs from the invasive woolly adelgid, or it might fall upon a broad-leafed black oak, flourishing despite drought; it might encounter a struggling scarlet tanager or a recently arrived red-bellied woodpecker. Some of these glimpses illustrate species in decline while others show them thriving, but none alone offers a full picture of the forest. Not only because they are limited in scope, but because they cannot show change as it unfolds. Cannot show that warmer winters help the woolly adelgid expand its range, which causes a decrease in hemlock populations; cannot show that red-bellied woodpeckers are moving north to track their thermal niche, while the number of scarlet tanagers just keeps dropping. The only way to paint the fullest picture possible is to hold each individual observation next to another, again and again, over time, until a story emerges.

This is, in many ways, what our interdisciplinary cruise—and, more broadly, the International Thwaites Glacier Collaboration—was designed to do. Here, then, are some of the things we saw, findings which have since been published in various scientific journals: Deep, interconnected troughs, up to 40 percent larger than predicted, in which thick bands of warm water are working their way under the ice; a clockwise circulation of this warm current, pulsing through Pine Island Bay and pushing under Thwaites, indicating that the water in these troughs mixes beneath the ice shelf and eats away at what remains from all directions. Aerial evidence of growing cracks in the only remaining ice shelf buttressing Thwaites, implying that it, too, will soon break apart, perhaps even in the next five years. Penguin bones that tell us the ice shelf didn't regenerate—not even once—over the last ten thousand years, making it unlikely to do so in the near future. And, perhaps most importantly, corrugation ridges that make us think that sometime during the last couple hundred years, and maybe even as recently as the middle of the twentieth century, Thwaites retreated at rates two to three times faster than we have ever witnessed—indicating that this glacier (and the many other marine-terminating glaciers like it) have the potential to break apart much more quickly than previously imagined.

Each of these observations helps us to better understand Thwaites and the systems that impact the glacier's current behavior, but even when we stitch them together and ask what kind of narrative emerges, we see just how fragmentary our knowledge is, how hard it is to make it coalesce into a single story. Even if we could spend the rest of our lives near the ice, complete comprehension wouldn't be possible. Not even close. After all, a significant chunk of Thwaites's ice shelf fell apart all around us, and we barely noticed.

In my limited lifetime—I am not yet forty—climate change has gone from something that we thought would happen in the future, to something that is happening now, to something that is accelerating at such a surprising pace that it makes most of our attempts to reckon with it outdated before they even get underway. Why shouldn't this great acceleration reach Antarctica? We saw some change while we were there, but, as I write now, I sense that what is to come will outpace anything we witnessed, and exponentially. I suspect that what we saw will seem sweet, almost harmless, when compared with what will unfold later in this century and beyond.

Paleoclimate records, like those the mud contains, hold clues to ice sheet insecurity during climate regimes far different from the one we live in today. As modelers fit these kinds of data together with present-day ice sheet behavior, uncertainty in sea level–rise projections is temporarily increasing, simply because so much has been missing from these projections in the past. Regardless of this increased uncertainty in the short term, there are still steps we can take while we wait for the improved models that will eventually arrive as result of all we are learning.

We must, in this unsettled window of time, continue with the difficult labor of living in and imagining a changed world, one that can become more equitable as a result of the climate crisis, not less. What will happen in 2100 is uncertain, but what is happening all along our coastline at the current moment is not. Our lowest-lying (and often low-income) communities are already suffering as a result of rising seas. Which means we need to develop and implement a practical and flexible framework for just climate adaptation now. This framework must center the people and places hurt by half a millennium of extraction, people and places that have not, as of yet, received proper recompense for all they have lost along the way. Then, as the numbers become clearer, we will already have in place the laws and policies that will help us meet the challenges we face. Because real climate resilience is something we have either together or not at all. These kinds of actions can also help to tackle the existential crisis that climate change presents and, as such, are much more useful than continuing to worry about all we do not know. Because if I have learned anything, it's this: Thwaites, and the whole of which it is part, will always be something we only partially understand.

•

I HAVE KEPT TRACK, SINCE the end of the cruise, of some of the people whom I lived alongside during those bright months, albeit only loosely. I saw Scott and Meghan and Joee in Waterville, Maine, when I gave a talk for the Sierra Club. Afterward we stood in the fluorescent-lit foyer and caught up: Joee quit her job as a marine technician, wanting to focus on conservation closer to home; Scott's partner joined him in Maine;

and Meghan decided to pursue a PhD. Rob's daughter graduated from college, and Bastien moved to Sweden to join Anna and her team at the University of Gothenburg. He also went on a tandem bike tour in Trier, Germany, with the woman he married. Barry and I ate lunch at a vegetarian place in Providence the fall after the cruise, and he told me he was looking forward to returning to Thwaites. Which he did. As did Ali, Rob, Rachel, Kelly, and a few others. But they weren't able to get anywhere near as close to the glacier as we did, due to unfavorable sea-ice conditions.

On that cruise, the one that followed ours, those onboard watched COVID-19 unfold at a distance, as they worked in perfect isolation. When it was time to head back home, the Brits were routed through Rothera Base and then the Falklands. Ali and his family got stranded in the UK for months. They rented an Airbnb, made the best of it. The crew spent a week off the coast of Chile, waiting for the borders to open again. Becky worked on her tenure case while homeschooling three boys. In May, she sent me photos of them doing yoga in their living room. Joee sent photos of frozen lakes and wind-stripped trees. Lindsey still works for the Antarctic Support Contract as a marine project coordinator, though she and Carmen don't deploy together anymore. The first summer of the pandemic, for fun, they sailed the Strait of Juan de Fuca with their son. In the photos Lindsey sent, Anvers has bright red hair and unsurprisingly looks right at home on a boat. Occasionally, Rachel and I comment on each other's Twitter posts, but other than that we have mostly lost touch. However, when she gave a presentation on her PhD work at the annual meeting of scientists studying the West Antarctica Ice Sheet, pulling together information from four different cores gathered on our cruise, I attended virtually and took notes and wished her well at the end. As for the seals that Lars and the rest tagged, they sent back 4,264 vertical profiles of temperature and salinity across the inner Amundsen over the year after we left.

George worked at the South Pole one season, then was back on the *Palmer.* Jack still works for Chouest, but instead of deploying on icebreakers to Antarctica, he cooks for the rocket scientists at Elon Musk's misguided SpaceX program. Jermaine lost his father, who worked on the *Palmer* for decades, to lung cancer shortly after the pandemic started. His daughter had just been born, which helped ease the grief. Luke has also become a dad since returning from Thwaites. When we last spoke, he was back home in North Carolina, between cruises. And whenever I have questions about the *Palmer*'s radar or onboard library, I contact Rick, who always responds with the information I've asked for and with a photo or two of the view from the bridge. Sometimes it's icebergs, sometimes it's the dry dock in Talcahuano, Chile. I have not caught up with Julian or Fernando or Captain Brandon—although I sent each of them a copy of the book in progress and asked for feedback, which they gladly gave.

The second year of the pandemic, all offshore research in Antarctica halted. Then it began again. As I type, Gui and Anna and Lars are all in quarantine, preparing to sail south on the *Palmer*. They're posting photos of the birds they can spot from their hotel windows and of the makeshift desks they've constructed. Which means that soon some of us will be back together again. Soon some of us will try to return to the ice that set this entire project in motion. When the biennial Thwaites Glacier research conference resumes, perhaps I, too, will go. More of us will be reunited there, but we will be far away from the thing that brought us together in the first place. Until then, the pages of this book keep us close, but in a superficial way. We are now as we were before: sometimes working toward each other, sometimes alongside each other, but mostly apart. To suggest otherwise is to mistake longing for the truth.

Also, this: I gave birth. For the first year of Nico's life, we rarely ventured farther than twenty miles from home. When he got big enough, we biked together: him in his toddler seat, barely supporting the weight of his own head, me panting and pedaling and describing the world. Instead of saying, "Look at the bay, look at the trees," I tried and continue to try changing who acts in my sentences. "The tupelo trees are greeting you; the Narragansett Bay swims by." Maybe if the language I use understands the landscape where our lives unfold as animate, he will too. I'll admit that, at first, this felt like a thought experiment, and not one that came easily to me. But just a couple of days ago, I was walking on the beach and realized I didn't have to try anymore to think about this ocean as a person: it just was and is.

Notes

xi *__You can't count how much__* Fred Moten, in Fred Moten and Stefano Harney, *The Undercommons: Fugitive Planning and Black Study* (New York: Minor Compositions, 2013), 154.

CAST OF CHARACTERS

1 *__Fifty-seven people sail to Thwaites Glacier__* Between January 29 and March 25, 2019, I interviewed many of my shipmates on the *Nathaniel B. Palmer* not only about their time on the ship but also about their birth experiences and what it means to choose to make more life (or not) as the planet's operating system tilts out of balance. I then transcribed all 213 of these interviews, creating an original archive that chronicles, in real time, this unprecedented expedition from the perspective of its scientists and crew. Anytime someone speaks outside of a scene in this book—in what I call testimonials—the words they use come from these interview transcriptions. Every single person who delivers a monologue in this book has reviewed, and in some cases fine-tuned, the speech ascribed to them. Finally, it is important to note that I occasionally used some of this material to reconstruct a scene, especially those where I was working in the field alongside the scientists—moments where I prioritized being available to help, and so might have tucked away my notebook or recording device.

1 *__The scientists belong to three teams__* If you are interested in learning more about the experiments being conducted as part of the International Thwaites Glacier Collaboration (ITGC), see the ITGC's official website: https://thwaitesglacier.org/.

1 *__The GHC and THOR teams__* Each of the acronyms for the scientific teams provide descriptions of the kinds of work the scientists

perform or of the kinds of questions they ask. For instance, GHC stands for Geological History Constraints on the Magnitude of Grounding-Line Retreat in the Thwaites Glacier System. THOR stands for Thwaites Offshore Research, and TARSAN stands for Thwaites-Amundsen Regional Survey and Network Integrating Atmosphere-Ice-Ocean Processes.

PROLOGUE

7 ***My due date was May 25th*** I recorded this conversation with my mother in early March 2020. I was six months pregnant at the time. Soon our worlds would turn upside down, as COVID fundamentally restructured every aspect of our daily lives.

DEPARTURES

15 ***the jellyfish heap up on the shore*** To learn more about mass jellyfish die-offs, see Timothy Jones et al., "Long-term Patterns of Mass Stranding of the Colonial Cnidarian *Velella velella*: Influence of Environmental Forcing," *Marine Ecology Progress Series* 662 (2021): 69–83, https://doi.org/10.3354/meps13644.

15 ***the pollinator plants just keep blooming*** Over the past couple decades, warmer temperatures have driven earlier pollen-season start dates and later pollen-season end dates. This study provides a good literature review of what we know about climate change and pollination while also attempting to project the potential future of pollination: Yingxiao Zhang and Allison L. Steiner, "Projected Climate-Driven Changes in Pollen Emission Season Length and Magnitude over the Continental United States," *Nature Communications* 13, (2022), https://doi.org/10.1038/s41467-022-28764-0.

15 ***long after the monarchs ought to be gone*** Monarchs begin to head south based on the angle of the sun, though there is speculation that timing and occurrence of pollination and monarch survivorship are related and negatively impacted by climate change. See Orley R. Taylor Jr. et al., "Is the Timing, Pace, and Success of the Monarch Migration Associated with Sun Angle?," *Frontiers in Ecology and Evolution* 7 (December 2019), https://doi.org/10.3389/fevo.2019.00442.

15 ***another year in which the ice lets go*** Calculating just how much ice Antarctica is losing is difficult. Most estimates depend upon satellite data to calculate the remaining mass of the continent's ice sheets. This data only began to be produced in 1979, and the earliest

data offers incomplete coverage. (See Caitlyn Kennedy, "Earliest Satellite Images of Antarctica Reveal Highs and Lows for Sea Ice in the 1960s," Climate.gov, November 4, 2014, https://www.climate.gov/news-features/featured-images/earliest-satellite-images-antarctica-reveal-highs-and-lows-sea-ice.) As the total amount of observational data has increased, so too have the methods for calculating ice loss, thus estimates of the total loss vary. However, scientists agree that West Antarctica is losing ice much more rapidly than other sectors of the continent. For more, see these two studies that give a range for ice loss: Eric Rignot et al., "Four Decades of Antarctic Ice Sheet Mass Balance from 1979–2017," *Proceedings of the National Academy of Sciences* 116, no. 4 (January 2019): 1095–1103, https://doi.org/10.1073/pnas.1812883116; and Isabella Velicogna et al., "Continuity of Ice Sheet Mass Loss in Greenland and Antarctica from the GRACE and GRACE Follow-On Missions," *Geophysical Research Letters* 47, no. 8 (April 2020), https://doi.org/10.1029/2020GL087291.

17 ***I visited with hundreds of flood survivors*** If you are interested in learning about climate change's early impact on coastal communities all around the United States, see my previous book, *Rising: Dispatches from the New American Shore* (Minneapolis, MN: Milkweed Editions, 2018).

17 ***I read an article about Thwaites*** This article was written by Jeff Goodell, who also deployed to Thwaites on the *Nathaniel B. Palmer*: Jeff Goodell, "The Doomsday Glacier," *Rolling Stone*, May 9, 2017, https://www.rollingstone.com/politics/politics-features/the-doomsday-glacier-113792/. In some ways, he is responsible for my seeking out the opportunity to be a part of this mission, and he logged some riveting dispatches from the ship. You can find them filed under "Journey to Antarctica," beginning with Jeff Goodell, "Journey to Antarctica: Jeff Goodell Begins His Trip to Thwaites Glacier," *Rolling Stone*, January 30, 2019, https://www.rollingstone.com/politics/politics-features/jeff-goodell-journey-to-antarctica-dispatch-1-786538/.

17 ***over two feet of potential sea level rise*** International Thwaites Glacier Collaboration, "Everything You Ever Wanted to Know about Thwaites Glacier & ITGC," June 2020, https://thwaitesglacier.org/sites/default/files/2022-01/ThwaitesGlacierFacts_UpdateJan2022.pdf.

17 ***Doomsday Glacier*** Goodell, "Doomsday Glacier."

17 ***no one has ever before been to Thwaites's calving edge*** Others have stood atop Thwaites: the first person to do so was Charles Bentley,

in the late 1950s. But the place where the ice shelf meets the ocean is incredibly difficult to reach due to sea ice, and the people on our cruise were the first to see it in the history of the planet. There would be two subsequent sea-based missions to Thwaites as part of the ITGC, but neither was able to get as close as we did during the initial mission in 2019.

18 ***a mixture of science and speculation*** For a great introduction to what we know, what we don't know, and what we suspect about Antarctica's ice sheets, see Richard B. Alley, "Are Antarctica's Glaciers Collapsing?," *Scientific American*, February 1, 2019, https://www.scientificamerican.com/article/are-antarcticas-glaciers-collapsing/.

18 ***our predictions about the speed of sea level rise*** The editor of the *Antarctic Sun* conducted a great conversation with Jeremy Bassis, one of the modelers at work on the ITGC, that explains in plain language what we do and don't know about sea level rise and Marine Ice Cliff Instability (MICI): Michael Lucibella, "Thwaites Glacier—Future," *Antarctic Sun*, May 17, 2021, https://antarcticsun.usap.gov/science/4450/.

19 ***M—, and I wanted to have Í—*** Anytime someone delivers a testimony in this book and is speaking about a family member, or about another person who does not have a public or professional life associated with the questions raised in this book, I have chosen to omit most of their name, for privacy.

21 ***more like a pancake than a cerebrum*** For a good introduction to Antarctica sea ice coverage, see British Antarctic Survey, "Seasonal Change," Discovering Antarctica, December 12, 2015, https://discoveringantarctica.org.uk/oceans-atmosphere-landscape/a-changing-climate/seasonal-change/.

21 ***Carolyn Beeler, a reporter covering the cruise*** Like Jeff Goodell, Carolyn did interesting reporting while onboard the *Palmer*. You can check out all of her stories for Public Radio International here: Carolyn Beeler, "Into the Thaw: Decoding Thwaites Glacier," *The World*, July 1, 2019, https://theworld.org/categories/thaw-decoding-thwaites-glacier.

22 ***Erika Blumenfeld, a photographer*** Anyone looking at these notes should put down this book and go check out Erika's stunning photographs. Her work was also used to illustrate "First Passage," an essay I wrote about this expedition for *Orion* magazine, back in the summer of 2021. See: https://erikablumenfeld.com/works/the-polar-project/.

23 ***Pregnant people, I learned, are not allowed to sail south*** In order to participate in the United States Antarctic Program, you must pass a series of physical examinations, after which you are declared

"physically qualified" or "not physically qualified." Pregnancy is one of the reasons a person might be determined not physically qualified. (Other countries have different rules that dictate who does and does not get deployed to the ice.)

26 ***a figure that would have been all but unthinkable*** This information comes from Elizabeth Chipman, *Women on the Ice: A History of Women in the Far South* (Victoria: Melbourne University Press, 1986), which I discuss at length in a subsequent citation. It should also be noted that there has been, over the past two decades, considerable effort to increase the number of women working on the ice. Erin Pettit, a professor at Oregon State University and a member of the ITGC, founded Inspiring Girls Expeditions and Girls on Ice, both of which provide tuition-free opportunities for young women to explore the cryosphere through field experience, science, and art: https://www.inspiringgirls.org/.

27 ***'We' are why we are here*** Jerri Nielsen and Maryanne Vollers, *Ice Bound: A Doctor's Incredible Battle for Survival at the South Pole* (New York: Talk Miramax Books/Hyperion, 2001), 216.

31 ***Aristotle imagined the earth as a sphere*** Robert Clancy et al., *Mapping Antarctica: A Five Hundred Year Record of Discovery* (Chichester, UK: Springer and Praxis Publishing, 2014), 67.

31 ***Ptolemy reasoned*** Joy McCann, *Wild Sea: A History of the Southern Ocean* (Chicago: University of Chicago Press, 2019), 10.

31 ***The name itself didn't arrive*** Bernadette Hince, *The Antarctic Dictionary: A Complete Guide to Antarctic English* (Collingwood, Australia: CSIRO Publishing, 2000), 7.

31 **antarktikós** *Merriam-Webster*, s.v. "Antarctica (*n.*)," accessed August 30, 2022, https://www.etymonline.com/word/antarctic.

31 ***first appeared on world maps in 1508*** Clancy et al., *Mapping Antarctica*, 61.

31 ***one-fourth of the total area of earth*** See Abraham Ortelius's map, *Typus orbis terrarum*, which was included in the world's first atlas. A reproduction can be accessed at the website of the Library of Congress, https://www.loc.gov/resource/g3200.ct007008/?r=-0.455, -0.002,1.864,0.874,0.

32 ***Some say Nathaniel Brown Palmer*** There are three primary claimants for the first sighting of Antarctica. David Day provides further detail on each case in his encyclopedic book, *Antarctica: A Biography* (New York: Oxford University Press, 2013). It is less clear who is the first to make it to Antarctic waters. Many claim Captain James Cook is the first to cross the Antarctic Circle, though recent research also suggests that it is

possible that a famous Māori explorer, Hui Te Rangiora, may have made it that far in the early seventh century. See: Priscilla M. Wehi et al., "A short scan of Māori journeys to Antarctica," *Journal of the Royal Society of New Zealand,* 52 (June 2021): 587–598, https://doi-org.revproxy.brown.edu/10.1080/03036758.2021.1917633.

32 ***Open an encyclopedia of Antarctic place names*** The single largest gazetteer of Antarctic place names in English that I've found was printed by the United States Geological Survey (USGS): United States Board on Geographic Names, *Geographic Names of the Antarctic*, ed. Fred G. Alberts, 2nd ed. (Reston, VA: USGS, 1995), https://pubs.er.usgs.gov/publication/70039167.

33 ***Men had been quarrelling*** Sara Wheeler, *Terra Incognita: Travels in Antarctica* (New York: Random House, 1996), xviii.

33 ***These are the origin stories*** Consider taking a look at Elizabeth Leane's thought-provoking and thorough article, "The Antarctic in Literature and the Popular Imagination," published in Mark Nuttall, Torben R. Christensen, and Martin J. Siegert, eds., *The Routledge Handbook of the Polar Regions* (London: Routledge, 2018).

36 ***Whales?*** Tasha Snow, the media coordinator, wrote a blog about the journey called "Snow on Ice." She writes about some of the same ship experiences as I do but from a slightly different perspective. Read about seeing the whales here: Tasha Snow, "Snow on Ice: Setting Sail #2," ITGC, January 30, 2019, https://thwaitesglacier.org/blog/snow-ice-setting-sail-2.

38 ***Objectivity is valued above all else*** Many feminist scholars have investigated the link between gender and the value ascribed to objectivity in the sciences. Here are two works of particular importance to me: Carolyn Merchant, *The Death of Nature: Women, Ecology, and the Scientific Revolution* (San Francisco: HarperCollins, 2019); and Donna Haraway, *Primate Visions: Gender, Race, and Nature in the World of Modern Science* (London: Routledge, 2015). As for the question of whether a writer can ever be objective, this is unsurprisingly something writers in particular like to debate. I personally recommend, on this subject, John D'Agata and Jim Fingal's *The Lifespan of a Fact* (New York: W. W. Norton, 2012).

38 ***The common name (pronounced* sigh*)*** These lines come from the guidebook referenced below. I have cut out some of the information in the entry for readability. Randall R. Reeves, *The National Audubon Society Guide to Marine Mammals of the World.* (New York: Alfred A. Knopf, 2008), 226.

39 ***The continent is commonly compared to a virgin*** This information comes from Elizabeth Leane, "Placing Women in the Antarctic Literary Landscape," *Signs: Journal of Women in Culture and Society* 34, no. 3 (Spring 2009): 509–14, https://doi.org/10.1086/593347.

40 ***broad white bosom*** Ernest Shackleton, "To the Great Ice Barrier," quoted in Elizabeth Leane, *Antarctica in Fiction: Imaginative Narratives of the Far South* (Cambridge, UK: Cambridge University Press, 2012), 59.

40 ***impenetrable*** Frederick A. Cook, *Through the First Antarctic Night, 1898–1899* (New York: Doubleday & McClure, 1900), 283.

40 ***women weren't welcome*** Elizabeth Chipman has written easily the most detailed account of women in Antarctica, *Women on the Ice*, which served as an invaluable resource during my research. As she notes, the main exception to the rule against women deploying through government-run programs in the 1950s was Russia. Since women were frequently present on Russian whaling vessels, there was less reluctance to include them in the state-sponsored scientific projects that would define Antarctica throughout the 1950s and '60s. It is also important to mention that the United States Antarctic Program began, in earnest, in 1959. I had expected that the government's involvement would act as a kind of open sesame for women who wished to venture south, but found the opposite was true. The Navy handled the logistical operations of research stations and vessels for much of the twentieth century, significantly limiting women's direct participation in the program. Rear Admiral James Reedy, the Commander of the United States Naval Support Force in Antarctica in the 1960s, infamously referred to it as "the womanless white continent of peace" (quoted in Chipman, 87). When the first all-women scientific expedition finally landed at the South Pole later that decade, journalist Walter Sullivan described the undertaking as "an incursion of females" into "the largest male sanctuary remaining on this planet" (Walter Sullivan, "Antarctic, a No-Woman's Land, to Get 6 Females," *New York Times*, October 1, 1969). I am also drawing, in this section, from Lisa Bloom's *Gender on Ice: American Ideologies of Polar Expeditions* (Minneapolis, MN: University of Minnesota Press, 1993). Finally, I would be remiss if I didn't mention this influential article on the impact of gender imbalance on the very study of glaciers: Mark Carey et al., "Glaciers, Gender, and Science: A Feminist Glaciology Framework for Global Environmental Change Research" *Progress*

in Human Geography 40, no. 6 (December 2016): 770–93, https://doi.org/10.1177/0309132515623368. It played a significant role in shaping my thinking on this subject.

40 ***While people of color occasionally made it*** As a recent study shows, racial minorities are systematically excluded from geology and climate change science: see Natasha Dowey et al., "A UK Perspective on Tackling the Geoscience Racial Diversity Crisis in the Global North," *Nature Geoscience* 14 (2021): 256–59, https://doi.org/10.1038/s41561-021-00737-w. This certainly impacts not just who studies what and how but the very questions we are asking about what is happening in Antarctica today. I worked with the Antarctic Support Contract to figure out if Julian was the first person from Jamaica to reach McMurdo, and the answer was "probably not"—but it is next to impossible to trace the countries of origin of the service employees in Antarctica. Probably one of the best books written about race and Antarctica is Mat Johnson's satirical novel *Pym* (New York: Spiegel & Grau, 2011). *Pym* follows a down-on-his-luck academic who studies the "pathology of Whiteness" (14) in Edgar Allan Poe's only novel, *The Narrative of Arthur Gordon Pym of Nantucket,* but who has not received tenure because he refuses to serve on his university's diversity and inclusion committee. He sets out for the South Pole with an all-black expedition crew half-interested in literature, half-interested in bottling glacial meltwater and selling it to yuppies back home. Come for the searing social commentary, stay for the abominable snowmen.

40 ***Boston University professor who had taunted*** About a month after my return, Boston University fired geologist David Marchant (Meredith Wadman, "Boston University Fires Geologist Found to Have Harassed Women in Antarctica," *Science*, April 12, 2019, https://www.science.org/content/article/boston-university-fires-geologist-who-sexually-harassed-women-antarctica). The glacier which bore his name had already been renamed Matataua, "a Maori word meaning 'a scout before the troops'" (Elizabeth Culotta. "Antarctic Glacier Gets New Name in Wake of Sexual Harassment Finding," *Science*, September 18, 2018, https://www.science.org/content/article/antarctica-glacier-gets-new-name-wake-sexual-harassment-finding).

41 ***Our bodies prime our metaphors*** James Geary, *I Is an Other: The Secret Life of Metaphor and How It Shapes the Way We See the World* (New York: HarperCollins, 2011), 100.

41 ***Intergovernmental Panel on Climate Change*** This report is still widely referenced when we talk about why meeting certain climate change mitigation benchmarks is so important. See the special report here: Intergovernmental Panel on Climate Change, "Global Warming of 1.5°C," (Cambridge, UK and New York, NY: Cambridge University Press, 2018), https://www.ipcc.ch/sr15/.

41 ***The problem with Antarctic storytelling*** Just before setting sail, I read about a present-day competition to claim another Antarctic first: to traverse its entirety alone and unaided. While we sailed to Thwaites, Colin O'Brady and Louis Rudd raced each other across the 921 miles that separate the Ronne Ice Shelf from the Ross Ice Shelf. According to the *New York Times*, "Both men hope to conquer a continent that has become the new Everest for extreme athletes" (Adam Skolnick, "No One Has Ever Crossed Antarctica Unsupported. Two Men Are Trying Right Now," *New York Times*, November 11, 2018, https://www.nytimes.com/2018/11/11/sports/antarctica-race.html). I'm not sure what irked me more when I read about their undertaking: the fact that any walk across Antarctica, no matter how solitary, is far from unsupported; or that such articles recycle stale metaphors that do little to lessen the distance between us and the ice. You might say it is not the old explorers I dislike, nor their lofty rhetoric around polar exploration, but that rhetoric's perverted persistence.

41 ***an actor in its own right*** Many environmental thinkers discuss the importance and difficulty (especially if you are raised within a culture where this is not the norm) of recognizing the animacy of the more-than-human world. Five writers' work immediately comes to mind: Robin Wall Kimmerer, *Braiding Sweetgrass: Indigenous Wisdom, Scientific Knowledge, and the Teachings of Plants* (Minneapolis, MN: Milkweed Editions, 2013); Amitav Ghosh, *The Nutmeg's Curse: Parables for a Planet in Crisis* (Chicago: University of Chicago Press, 2021); Anna Lowenhaupt Tsing, *The Mushroom at the End of the World: On the Possibility of Life in Capitalist Ruins* (Princeton: Princeton University Press, 2015); Julie Cruikshank, *Do Glaciers Listen?: Local Knowledge, Colonial Encounters, and Social Imagination* (Vancouver: University of British Columbia Press, 2005); and Bathsheba Demuth, *Floating Coast: An Environmental History of the Bering Strait* (New York: W. W. Norton, 2019).

STALLED

43 ***hundreds of thousands of gallons of diesel*** This might sound like a lot—and it is—but a Boeing 747 would burn the same amount circling the earth continuously for just five days, rather than the two months this was used to sustain the *Palmer*. What I mean to point to here is not the relative pros and cons of our usage but just how fuel intensive airplane travel is—and how deeply entwined our lives are with the fossil fuel industry.

49 ***Edison Chouest, the shrimper*** Chouest was important enough that the *New York Times* ran a substantial obituary when he died: Associated Press, "Edison Chouest, Shipyard Founder, Dies at '91," *New York Times*, October 3, 2008, https://www.nytimes.com/2008/10/04/business/04chouest.html.

49 ***the Louisiana of the present*** ProPublica has created one of the best immersive stories about land loss in Louisiana: Bob Marshall et al., "Losing Ground," ProPublica, August 28, 2014, https://projects.propublica.org/louisiana.

50 **Look at this iceberg** There is a lot of really asinine "news," also known as clickbait, about Antarctica. Right before I set out for Thwaites, there was a particularly frustrating flurry of stories about a "perfectly rectangular" iceberg. Here's one: Trevor Nace, "NASA Finds Perfectly Rectangular Iceberg in Antarctica as If It Was Deliberately Cut," *Forbes*, October 22, 2018, https://www.forbes.com/sites/trevornace/2018/10/22/nasa-finds-perfectly-rectangular-iceberg-in-antarctica-as-if-it-was-deliberately-cut/.

52 ***Ursula Le Guin's short story "Sur"*** The story was first published in the *New Yorker* on February 1, 1982. It would later be anthologized in numerous collections. I read it in *The Wide White Page,* which provides an excellent introduction to the many ways that Antarctica has been written about, starting with Dante: Bill Manhire, ed., *The Wide White Page: Writers Imagine Antarctica* (Wellington, New Zealand: Victoria University Press, 2004).

52 ***Although I have no intention*** Le Guin, "Sur," in *Wide White Page*, 90.

52 ***abominable seaport*** Le Guin, 94.

53 ***not very seriously*** Le Guin, 104.

53 ***some kind of mark*** Le Guin, 105.

53 ***Achievement is smaller*** Le Guin, 98.

53 ***hoarse as a skua*** Le Guin, 108.

56 ***a required safety training*** One of my shipmates, Linda Welzenbach of Rice University, maintained a detailed blog during our expedition. Many of the day-to-day details included in this book can be cross-referenced there, including this safety training. She is also a phenomenal photographer, and I suggest visiting her site just to get a stronger visual sense of this story. You can see all of her blog posts here: https://thwaitesglacieroffshoreresearch.org/blogs.

FIRST PASSAGE

64 ***the first people to send an autonomous vehicle*** *Nature* published an article on this aspect of the TARSAN project: Jeff Tollefson. "First Look under Imperilled Antarctic Glacier Finds 'Warm Water Coming from All Directions,'" *Nature*, February 20, 2020, https://www.nature.com/articles/d41586-020-00497-4. To learn more about how central these machines are to scientific advancement, check out Russell B. Wynn et al., "Autonomous Underwater Vehicles (AUVs): Their Past, Present and Future Contributions to the Advancement of Marine Geoscience," *Marine Geology* 352 (June 2014): 451–68, https://doi.org/10.1016/j.margeo.2014.03.012.

65 ***a laminated map of the inner Amundsen Sea*** Here is an entry in Linda Welzenbach's blog that shows Kelly and Ali looking over one of the maps covered in snail trails: "The 'Iceberg' Correction," *THOR Blogs* (blog), Thwaites Glacier Offshore Research (THOR), February 4, 2019, https://thwaitesglacieroffshoreresearch.org/blogs/2019/2/14/the-iceberg-correction.

66 ***Pine Island Glacier*** As early as 1981, T. J. Hughes, of the University of Maine, suggested that Thwaites and Pine Island might be "the weak underbelly of the West Antarctic Ice Sheet," their topography leading to potential collapse of much of the ice sheet that defines lesser Antarctica: T. J. Hughes, "The Weak Underbelly of the West Antarctic Ice Sheet," *Journal of Glaciology* 27, no. 97 (1981): 518–25, https://doi.org/10.3189/S002214300001159X. This hypothesis was formulated from the author's work with his colleague G. H. Denton, modeling past glacial maximums. Since then, Eric Rignot (among others) has performed some of the most extensive studies of the area, relying on data sets generated by satellites and aerial surveys. See Eric Rignot et al., "Four Decades of Antarctic Ice Sheet Mass Balance from 1979–2017," *Proceedings of the National Academy of Sciences* 116, no. 4 (January 2019): 1095–1103, https://doi.org/10.1073

/pnas.1812883116; and Eric Rignot et al., "Widespread, Rapid Grounding Line Retreat of Pine Island, Thwaites, Smith, and Kohler Glaciers, West Antarctica, from 1992 to 2011," *Geophysical Research Letters* 41, no. 10 (May 2014): 3502–09, https://doi.org/10.1002/2014GL060140. Finally, it should be noted that the scale at which these data are collected means that there are disagreements within the scientific community around just how much these glaciers have deteriorated during this time.

66 ***the ice sheet's death mask*** This is a reference to a paper written by one of the principal investigators (PIs) on the ITGC: J. S. Wellner et al., "The Death Mask of the Antarctic Ice Sheet: Comparison of Glacial Geomorphic Features across the Continental Shelf," *Geomorphology* 75, nos. 1–2 (April 2006): 157–71, https://doi.org/10.1016/j.geomorph.2005.05.015.

66 ***no significant topography underneath*** To learn more about the land beneath Thwaites, see John W. Holt et al., "New Boundary Conditions for the West Antarctic Ice Sheet: Subglacial Topography of the Thwaites and Smith Glacier Catchments," *Geophysical Research Letters* 33, no. 9 (May 2006), https://doi.org/10.1029/2005gl025561.

67 ***Alpine glaciers behave*** For a detailed discussion on all kinds of glaciers, ice caps, ice sheets, etc., see "Quick Facts, Basic Science, and Information about Snow, Ice, and Why the Cryosphere Matters," National Snow and Ice Data Center, https://nsidc.org/cryosphere/glaciers/questions/types.html.

67 ***Thwaites rests in a basin*** One of the best papers explaining why Thwaites is of such scientific concern is Ted Scambos et al., "How Much, How Fast?: A Science Review and Outlook for Research on the Instability of Antarctica's Thwaites Glacier in the 21st Century," *Global and Planetary Change* 153 (June 2017): 16–34, https://doi.org/10.1016/j.gloplacha.2017.04.008. Ted is one of the coordinating PIs on the ITGC, and this article is considered the foundation for all of the science they hope to conduct during the five years the ITGC is currently slated to run.

67 ***more than six times*** When I asked Rob where this number came from, he gave me a very specific and thoughtful answer, so I am going to transcribe it here: "This statistic is based on the numbers in Table 1 of the 2019 paper in *Proceedings of the National Academy of Sciences* by Eric Rignot et al. [which I have referenced elsewhere in these notes]. The table doesn't actually provide values for the net rate of ice loss directly, though. You can get them by subtracting the surface mass balance (SMB) number from the ice

discharge numbers, which are provided for different decades in different columns. If you do this simple calculation, you'll find that between 1979–1989, Thwaites's net loss rate was 4.6 gigatons per year, whereas between 2009–2017 it was 34.9 gigatons per year, which indicates a more than seven-fold net increase since the 1980s . . . but I'd prefer to stick with 'more than six times' to be conservative, given that there are some uncertainties in each of the individual measurements."

67 ***uncharacteristically ice free*** NASA has a neat little webpage that compares satellite images of Thwaites from 2001 to an image gathered about one month before our cruise departed from Punta Arenas: "Thwaites Glacier Transformed," NASA Earth Observatory, February 6, 2020, https://earthobservatory.nasa.gov/images/146247/thwaites-glacier-transformed. To see how much more of "Unnamed Bay" would open up while we were working in the area, see Linda Welzenbach's blog entry "Thwaites Glacier: There and Gone," *THOR Blogs* (blog), Thwaites Glacier Offshore Research (THOR), March 12, 2019, https://thwaitesglacieroffshoreresearch.org/blogs/2019/3/12/thwaites-glacier-there-and-gone.

68 ***eleven square* kilometers** I used the Shameplane site (which is now defunct) to calculate how our trip would impact sea-ice coverage. It should be noted that different carbon calculators give wildly varying results for the same flight due to variations in the underlying assumptions made in the calculator methodology.

68 ***all the things I do*** Jedediah Britton-Purdy has written a number of thoughtful meditations on how the kinds of human infrastructure our society develops and then depends upon determine so much of our carbon load. His articles also help me to embrace the idea that engineering another world is possible. Of particular importance are the pieces he has written for *Dissent* magazine—including, for example, "The World We've Built," *Dissent*, July 3, 2018, https://www.dissentmagazine.org/online_articles/world-we-built-sovereign-nature-infrastructure-leviathan.

69 ***When we set the size*** Elizabeth Kolbert, "The Case Against Kids," *New Yorker*, April 9, 2012, https://www.newyorker.com/magazine/2012/04/09/the-case-against-kids.

69 ***whose long roots can be traced*** See Jade S. Sasser, *On Infertile Ground: Population Control and Women's Rights in the Era of Climate Change* (New York: New York University Press, 2018).

69 ***To answer death with utopian futurity*** Alexis Pauline Gumbs, "m/other Ourselves: A Black Queer Feminist Genealogy for

Radical Mothering," in Alexis Pauline Gumbs, China Martens, and Mai'a Williams, eds., *Revolutionary Mothering: Love on the Front Lines* (Oakland, CA: PM Press, 2016), 21. Gumbs is writing from inside the robust reproductive justice canon, which centers mothers of color in the fight for a more egalitarian world. Her work is inspired by radical feminists like Audre Lorde and the SisterSong Collective.

69 ***the radical potential*** Gumbs, "m/other Ourselves," 21–23.

73 ***the haunting fear*** The answer that Barry and I didn't know is that this is H. L. Mencken, in *A Mencken Chrestomathy: His Own Selection of His Choicest Writings* (New York: Vintage, 1982), 624.

79 ***Crossing the Drake*** Joy McCann's *Wild Sea* is the single most accessible and exhaustive history of the Southern Ocean that I know of; it is well worth the read.

79 ***like an almond . . . in a chocolate bar*** Thomas Orde-Lees, quoted in Alfred Lansing, *Endurance: Shackleton's Incredible Voyage* (New York: Basic Books, 2014), 35.

80 ***We fought the seas and the winds*** Ernest Shackleton, *South: The Story of Shackleton's Last Expedition 1914–1917* (New York: Macmillan, 1926), 168. I groan about the Antarctic explorer stories in this book, but I'll admit that I did enjoy reading Shackleton's memoir.

80 ***my comrades who fell*** Shackleton, *South*, v.

80 ***two feet two inches*** This fabulous detail comes from Sara Wheeler's equally fabulous *Terra Incognita*.

84 ***What if, instead of carrying*** Ada Limón, "The Vulture & the Body," in *The Carrying* (Minneapolis, MN: Milkweed Editions, 2018), 13. At the time of our trip, this was Limón's most recent book, for which she was awarded the National Book Critics Circle Award. She has since published another excellent collection, *The Hurting Kind*. And she was named poet laureate of the United States; her poetry is *that* necessary and vital.

85 ***Simply put, we're living*** Shanna H. Swan and Stacey Colino, *Count Down: How Our Modern World Is Threatening Sperm Counts, Altering Male and Female Reproductive Development, and Imperiling the Future of the Human Race* (New York: Scribner, 2021), 2. Much of the information in this section comes from Swan and Colino's unsettling book.

85 ***women of color suffer higher rates*** Melissa F. Wellons et al., "Racial Differences in Self-Reported Infertility and Risk Factors for Infertility in a Cohort of Black and White Women: The CARDIA Women's Study," *Fertility and Sterility* 90, no. 5 (November 2008): 1640–48, https://doi.org/10.1016/j.fertnstert.2007.09.056.

85 ***red-tailed black cockatoo*** Stanley Mastrantonis et al., "Climate Change Indirectly Reduces Breeding Frequency of a Mobile Species through Changes in Food Availability," *Ecosphere* 10, no. 4 (April 2019), https://doi.org/10.1002/ecs2.2656.

85 ***temperature at the time of insemination*** Camryn Allen is my favorite person to talk to about sea turtle sex ratios and climate change. You can follow her on Twitter at @camryndallen. And you can read more about the phenomenon in this article: Lyndsey K. Tanabe et al., "Potential Feminization of Red Sea Turtle Hatchlings as Indicated by In Situ Sand Temperature Profiles," *Conservation Science and Practice* 2, no. 10 (October 2020), https://doi.org/10.1111/csp2.266. To read about some of the things humans have been doing to try to stave off the extinction of sea turtles, see Jana Blechschmidt et al., "Climate Change and Green Sea Turtle Sex Ratio—Preventing Possible Extinction," *Genes* 11, no. 5 (May 2020), https://doi.org/10.3390/genes11050588.

85 ***egg cells to appear in the testicles*** Tyrone Hayes is an endocrine disruption expert who is particularly taken with frogs. You can watch him speak here: Tyrone Hayes, "Endocrine Disruption, Environmental Justice, and the Ivory Tower: What the Privileged Need to Know," March 2018, TED video, 15:45, https://www.ted.com/talks/tyrone_hayes_endocrine_disruption_environmental_justice_and_the_ivory_tower.

85 ***polychlorinated biphenyls (or PCBs)*** Jean-Pierre Desforges et al., "Predicting Global Killer Whale Population Collapse from PCB Pollution," *Science* 361, no. 6409 (September 2018): 1373–76, https://doi.org/10.1126/science.aat1953.

87 ***They suspect that Thwaites*** Richard B. Alley et al., "Oceanic Forcing of Ice-Sheet Retreat: West Antarctica and More," *Annual Review of Earth and Planetary Sciences* 43, (May 2015): 207–31, https://doi.org/10.1146/annurev-earth-060614-105344.

88 ***Charles Bentley was one of the first people*** Much of this information comes from Jon Gertner's well-written introduction to Thwaites, "The Race to Understand Antarctica's Most Terrifying Glacier," *Wired*, December 10, 2018, https://www.wired.com/story/antarctica-thwaites-glacier-breaking-point/. I also drew on this obituary of Bentley: Michael Lucibella, "Charles Bentley 1929–2017," *Antarctic Sun*, August 30, 2017, https://antarcticsun.usap.gov/features/4323.

88 ***A few more images were taken*** This paper discusses how newly rediscovered archival film footage taken of Thwaites in the 1970s suggests that the ice is retreating even faster than we

tend to calculate today: Dustin Schroeder et al., "Multidecadal Observations of the Antarctic Ice Sheet from Restored Analog Radar Records," *Proceedings of the National Academy of Sciences* 116, no. 38 (September 2019): 18867–73, https://doi.org/10.1073/pnas.1821646116.

88 ***Any real calculations of the glacier's retreat*** Detailed, data-driven calculations of mass ice loss begin in the mid-1990s and are based on airborne data calculating snow accumulation and loss. See the IMBIE team, "Mass Balance of the Antarctic Ice Sheet from 1992 to 2017," *Nature* 558 (2018): 219–22, https://doi.org/10.1038/s41586-018-0179-y; and Brooke Medley et al., "Constraining the Recent Mass Balance of Pine Island and Thwaites Glaciers, West Antarctica, with Airborne Observations of Snow Accumulation," *Cryosphere* 8, no. 4 (2014): 1375–92, https://doi.org/10.5194/tc-8-1375-2014.

88 ***Modelers often use past events*** For more on storm modeling, see my and Rebecca Elliot's article "Stormy Waters," *Harper's*, June 2017, https://harpers.org/archive/2017/06/stormy-waters/.

89 ***the very processes*** This question, about what is physically happening to Antarctica's ice sheets, has been taken up by a handful of geologists over the past decade. Two different physical processes that might lead to rapid ice sheet disintegration have been proposed: Marine Ice Sheet Instability (MISI) and Marine Ice Cliff Instability (MICI), both of which might be at play with Thwaites. To learn more about them, I suggest the excellent article "How Much, How Fast?" by Ted Scambos et al. If you are interested in MICI specifically, see David Pollard et al., "Potential Antarctic Ice Sheet Retreat Driven by Hydrofracturing and Ice Cliff Failure," *Earth and Planetary Science Letters* 412 (February 2015): 112–21, https://doi.org/10.1016/j.epsl.2014.12.035. Since our return from Thwaites, there have been a couple important updates on the theories circulating around MICI and the likelihood that it will happen at Thwaites. The ITGC modeler Jeremy Bassis and a handful of other scientists have created dynamic models predicting that sea ice and other calved debris can slow (if only temporarily) the rate at which the glacier behind it flows. But if upstream ice thickens (as is the case at Thwaites), then catastrophic collapse can still occur. See Jeremy Bassis et al., "Transition to Marine Ice Cliff Instability Controlled by Ice Thickness Gradients and Velocity," *Science* 372, no. 6548 (June 2021): 1342–44, https://doi.org/10.1126/science.abf6271; and Anna J. Crawford et al., "Marine Ice-Cliff Instability Modeling Shows Mixed-Mode Ice-Cliff Failure and Yields Calving

Rate Parameterization," *Nature Communications* 12, (2021), https://doi.org/10.1038/s41467-021-23070-7.

89 ***We could be underestimating*** Richard B. Alley, "Are Antarctica's Glaciers Collapsing?," *Scientific American*, February 1, 2019, https://www.scientificamerican.com/article/are-antarcticas-glaciers-collapsing/.

89 ***these tiny corrugation ridges*** Alastair G. C. Graham et al., "Seabed Corrugations beneath an Antarctic Ice Shelf Revealed by Autonomous Underwater Vehicle Survey: Origin and Implications for the History of Pine Island Glacier," *Journal of Geophysical Research: Earth Surface* 118, no. 3 (September 2013): 1356–66, https://agupubs.onlinelibrary.wiley.com/doi/10.1002/jgrf.20087.

INTO THE ICE

99 ***Like a piston in a pump*** Michael P. Meredith, "The Global Importance of the Southern Ocean, and the Key Role of Its Freshwater Cycle," *Ocean Challenge: The Challenger Society for Marine Science* 23, no. 2 (2017): 27–32, https://nora.nerc.ac.uk/id/eprint/521441/.

103 ***Diana Richards*** I have changed this student's name.

104 ***the birth of the environmental movement*** It is interesting to me that there isn't (yet) a really critical, deeply thought-out history of the environmental movement in the United States. Most of the more surface-level investigations note the progression from the first Earth Day, which was all about safeguarding the planet for future generations, to now, where the environmental movement is largely youth-led; what they don't acknowledge is how prevalent the question of whether or not to have children has become among this younger demographic.

106 ***Ice is of direct concern*** Nathaniel Bowditch, *The American Practical Navigator: An Epitome of Navigation* (Bethesda, MD: Defense Mapping Agency Hydrographic/Topographic Center, 1995), 455.

113 ***It was BP [British Petroleum] that popularised*** Meehan Crist, "Is It OK to Have a Child?" *London Review of Books* 42, no. 5 (March 5, 2020), https://www.lrb.co.uk/the-paper/v42/n05/meehan-crist/is-it-ok-to-have-a-child. This article made a significant impact on my thinking about climate change and reproduction. Crist is currently expanding it into a book to be published with Random House.

113 ***the web page of an independent*** "BP," on K. J. Bowen's personal website, accessed September 7, 2022, https://kjbowen.com/clients/bp/.

113 ***What on earth*** "Beyond Petroleum" ad, 2005, on Bowen's personal website.

114 ***carbon footprint calculators*** At present, BP funds a carbon footprint calculator app that tracks how much CO_2 you pump into the air through your daily activities. Think of it as the FitBit of carbon calculation, as the journalist Kate Yoder urged: "Footprint Fantasy," *Grist*, August 26, 2020, https://grist.org/energy/footprint-fantasy/.

115 ***the BirthStrike movement*** There was loads of media coverage around the BirthStrike movement when it started, most of which got the founders' aims wrong. This is why I called Blythe Pepino and spoke to her myself. The BirthStrike movement has since disbanded, in part because of how the media hijacked its message. For more on this, see George Monbiot, "Population Panic Lets Rich People Off the Hook for the Climate Crisis They Are Fuelling," *The Guardian*, August 26, 2020, https://www.theguardian.com/commentisfree/2020/aug/26/panic-overpopulation-climate-crisis-consumption-environment.

115 ***we also reduce the output of*** CO_2 When we talk about how having one less child sequesters 60 tones of CO_2 it is vital to note that this number doesn't reflect the emissions that that hypothetical child would make over a single year; instead it adds together the potential emissions of one's future descendants and attributes (a portion of) them to the parent that chooses to have a child in the present tense. There has been significant back and forth within the scientific community (and beyond) about whether or not it is right or effective to hold individuals responsible in this way. Though there *are* studies that show that new parents tend to live more carbon intensive lives since time is a limited resource and they have to get more done with less of it. See: Jonas Nordström et al., "Do parents counter-balance the carbon emissions of their children?," *PLOS ONE* (April 15, 2021), https://doi.org/10.1371/journal.pone.0231105.

ISLANDS

122 ***Before sound became symbol*** This information comes from an email exchange I shared with Rob on August 27, 2019, when he wrote, "The only way we would know the age of the oldest ice on the Canisteo Peninsula for certain is if an ice core was drilled to the bed, and this has not been done. Inferences on the age of ice at different depths below surface in the region are based on ice flow

models informed by the nearest deep ice cores (WAIS divide) and climatology (surface temperature and surface accumulation distribution). Colleagues who have considered this with a view to selecting coastal ice coring sites tell me that there is unlikely to be any ice more than about four thousand years old around the coast of the Amundsen Sea. The reason there is no older ice is because it is a relatively high snow-accumulation-rate region, and this drives faster throughput."

122 ***The oldest ice cores ever drilled*** Older ice cores have been drilled, but we cannot date them in the same way, which makes using them as a proxy for past CO_2 levels more difficult. See Paul Voosen, "Record-Shattering 2.7-Million-Year-Old Ice Core Reveals Start of the Ice Ages," *Science*, August 15, 2017, https://www.science.org/content/article/record-shattering-27-million-year-old-ice-core-reveals-start-ice-ages.

124 ***Elders who know them well*** Cruikshank, *Do Glaciers Listen?*, 3. I was introduced to Cruikshank's work by Bathsheba Demuth, a friend and colleague at Brown University. Like Cruikshank, Demuth has spent many years above the Arctic Circle, studying how different cultures make sense of the scarcity and abundance that pulses through that place. As I worked on this project, I often thought of Bathsheba's book, *Floating Coast*, where she argues that bowhead whales—who for millennia lived alongside Indigenous hunters in the Bering Strait—chose to flee north, under the icepack, when the commercial fishing industry arrived in the middle of the nineteenth century. *We will not die that way, not for you*, their behavior seemed to say. Her ability to center the decisions of nonhuman actors comes from immense amounts of archival research, gathering of firsthand testimonies, and years spent paying attention to the animals and to the ice.

124 ***Between eighteen thousand and six thousand years ago*** There are lots of different ways to estimate past sea level–rise rates. This paper provides a good overview, compiling the findings from many types of data: Benjamin P. Horton et al., "Mapping Sea-Level Change in Time, Space, and Probability," *Geological Society of America Abstracts with Programs* 50, no. 6 (November 2018), https://doi.org/10.1130/abs/2018am-318319.

125 ***Were they not so eager to steal land and enslave*** Some of this thinking about just when the Anthropocene begins and its connection to colonialism and anti-Blackness arises out of my reading of Kathryn Yusoff's brilliant book *A Billion Black Anthropocences or None* (Minneapolis, MN: University of Minnesota Press, 2018).

126 ***Suddenly, everything seemed to click*** Andri Snær Magnason, *On Time and Water*, trans. Lytton Smith (Rochester, NY: Open Letter, 2021), 86.

129 **Folklore and the Sea** The book that Joee recommends is Horace Beck, *Folklore and the Sea* (Middletown, CT: Wesleyan University Press, 1973).

132 ***I mention an article*** Corey J. A. Bradshaw and Barry W. Brook, "Human Population Reduction Is Not a Quick Fix for Environmental Problems," *Proceedings of the National Academy of Sciences* 111, no. 46 (November 2014): 16610–15, https://doi.org/10.1073/pnas.1410465111.

133 ***"Objectivity or Heroism?"*** Naomi Oreskes, "Objectivity or Heroism?: On the Invisibility of Women in Science," *Osiris* 11 (1996): 87–113, https://www.jstor.org/stable/301928.

146 ***creates an optimal environment*** "Pre-Seed Fertility-Friendly Personal Lubricant," First Response, accessed September 8, 2022, https://firstresponsefertility.com/our-products/pre-seed.

151 ***Over fifty-six million years ago*** Johann P. Klages et al., "Temperate Rainforests near the South Pole during Peak Cretaceous Warmth," *Nature* 580 (April 2020): 81–86, https://doi.org/10.1038/s41586-020-2148-5.

151 ***CO_2 levels were one thousand parts per million*** "Is the Current Level of Atmospheric CO_2 Concentration Unprecedented in Earth's History?" Royal Society, accessed August 31, 2022, https://royalsociety.org/topics-policy/projects/climate-change-evidence-causes/question-7/.

154 ***the earth beneath it rebounds*** As I fact-checked this book, Scott reminded me of something that is particularly important when talking about rebound rates in the Amundsen Sea. He said, "If we are talking about the earth rebounding like a memory-foam mattress, it is important to note that not all mattresses are created equal. Some reform to their original shape much more quickly than others. That's the case in this part of Antarctica. Although sea level may be on the rise globally, at a local scale it appears to be falling, because the uplift of the islands after the glacier retreats outpaces changes in sea level."

154 ***Canisteo glacier*** While everyone on board referred to the glacier on top of the Canisteo Peninsula as the Canisteo glacier, technically what we see from the ship is called the Canisteo Peninsula ice front. Please see the paper generated from the fieldwork we conducted for a detailed map of our study area: Scott Braddock et al., "Relative Sea-Level Data Preclude Major Late Holocene

Ice-Mass Change in Pine Island Bay," *Nature Geoscience* 15, (July 2022): 568–72, https://doi.org/10.1038/s41561-022-00961-y.

BETWEEN THE PAST AND THE FUTURE

167 ***it would jeopardize*** Jennie Darlington, *My Antarctic Honeymoon: A Year at the Bottom of the World* (New York: Doubleday, 1957), 93.

167 ***There are some things*** Darlington, *My Antarctic Honeymoon*, 94.

167 ***"intuitively"*** Darlington, 93.

167 ***Inside me something hardened*** Darlington, 91.

167 ***"manhauling"*** Darlington, 162.

167 ***more for slamming purposes*** Darlington, 153.

168 ***The Antarctic had equalized*** Darlington, 272.

168 ***Before the icebreakers*** Darlington, 277.

183 ***Conceivable Future*** This is one of the most compelling groups to arise at the intersection of climate change and reproduction. You should 100 percent spend at least thirty minutes listening to the testimonies gathered on their website: https://conceivablefuture.org/.

184 ***hotter and less stable*** For a reconstruction of the earth's atmosphere over a good part of the time modern humans have existed, see Jeff Tollefson, "Earth Is Warmer Than It's Been in 125,000 Years," *Scientific American*, August 9, 2021, https://www.scientificamerican.com/article/earth-is-warmer-than-its-been-in-125-000-years/.

184 ***preparing to run for state senate*** Meghan Kallman won this race and has since contributed to some pretty amazing stuff in Rhode Island, including working on an affordable housing bill that is considered one of the most progressive in the country, starting a pilot program to make our public transportation fare free, and continuing the ongoing labor that goes into the safeguarding of abortion rights.

185 ***The Brits sent these seemingly silent pulses*** David Hambling, "7 Scientific Advances That Came out of World War I," *Popular Mechanics*, April 6, 2017, https://www.popularmechanics.com/military/research/g1577/7-surprising-scientific-advances-that-came-out-of-world-war-i/.

185 ***the first obstetric ultrasound*** Nicole Erjavic, "Ian Donald (1910–1987)," Embryo Project Encyclopedia, January 30, 2018, https://embryo.asu.edu/pages/ian-donald-1910-1987.

185 ***There is not so much difference*** Ian Donald, "On Launching a New Diagnostic Science," *American Journal of Obstetrics and Gynecology* 103, no. 5 (March 1969): 609–28, https://doi.org/10.1016/0002-9378(69)90559-6.

185 ***when the wave encounters something solid*** Kelly Hogan created an exceptional video introduction to the ins and outs of mapping the seafloor in Antarctica. It was made during the pandemic and is designed to help parents homeschool their kids, but it is also approachable and fascinating, even for adults: Kelly Hogan in "Mapping Deep Channels under Thwaites Glacier—Dr. Kelly Hogan," British Antarctic Survey, YouTube, March 27, 2021, video, 57:01, https://www.youtube.com/watch?v=jp6MDocLb0c.

190 ***her research into the effects of climate change*** Clarke is still in the process of publishing the results from this work, though some of her earlier work supports some of these claims. See Melody S. Clark et al., "Hypoxia impacts large adults first: consequences in a warming world," *Global Change Biology* 7, no. 37 (March 2013), https://doi.org/10.1111/gcb.12197.

190 ***adaptation through evolution*** To learn more about how climate change and evolution interact with one another, see Morgan Kelly, "Adaptation to Climate Change through Genetic Accommodation and Assimilation of Plastic Phenotypes," *Philosophical Transactions of the Royal Society B: Biological Sciences* 374, no. 1768 (March 2019), https://doi.org/10.1098/rstb.2018.0176.

192 ***the first women to spend a dark season at Rothera*** The British Antarctic Survey keeps track of everyone who has overwintered on the ice in a British Research Station here: "Database of Winterers," British Antarctic Survey Club, accessed September 7, 2022, https://basclub.org/winterers/.

193 ***rather than coerced*** As Maggie Nelson points out in her book *On Freedom*—see the next note—generosity and compulsion are more difficult to tease apart than we'd probably like to admit. This becomes obvious when you think about a parent's relationship to their child.

193 ***is the unceasing, rigorous work*** Maggie Nelson, *On Freedom: Four Songs of Care and Constraint* (Minneapolis, MN: Graywolf Press, 2021), 52.

ARRIVAL

200 ***its hemorrhaging heart of milk*** This line is inspired by Joan Naviyuk Kane's eponymous poem in her stunning collection *Hyperboreal* (Pittsburgh, PA: University of Pittsburgh Press, 2013).

203 ***a series of time-lapse photos*** You can see the video I created from these photos and others I shot on that first day here: Elizabeth Rush, "Here's What Antarctica's Calving Glaciers

Look Like Up Close," *National Geographic*, March 12, 2019, https://www.nationalgeographic.com/environment/article/watching-thwaites-glacier-calving-antarctica.

203 ***Seeing comes before words*** John Berger, *Ways of Seeing* (London, UK: Penguin, 1972), 7.

207 ***poetic exploration of beech trees*** C. D. Wright, *Casting Deep Shade: An Amble Inscribed to Beech Trees & Co* (Port Townsend, WA: Copper Canyon Press, 2019), 93.

207 ***beech trees (one of the many species*** Howard Falcon-Long, "Secrets of Antarctica's Fossilized Forests," *BBC*, February 8, 2011, https://www.bbc.com/news/science-environment-12378934.

207 ***the largest climate protest in history*** These climate protests were inspired, in part, by Greta Thunberg and organized to occur in the lead-up to the United Nations' climate summit. The energy around these strikes was intoxicating, and it was clear that the climate movement was gaining a lot of ground. The pandemic—both the suggestions to avoid large crowds and the way in which it demanded so much of our attention—helped to dampen the momentum around the youth-led climate uprising in the months that followed. That being said, the Inflation Reduction Act was passed in 2022 and marks the single largest piece of legislation to address climate change in the United States' history. It isn't enough, but it is a start.

208 ***discuss the letter we're drafting to Governor Raimondo*** Our organization played a role in overturning the governor's nomination, which is explained in more detail here: Patricia Burke, "Raimondo's PUC Pick Needs Scrutiny for National Grid Connections," Uprise RI, June 9, 2019, https://upriseri.com/2019-06-09-puc/.

210 ***to generate a profile*** Some of the information the glider collected was incorporated into Anna Wåhlin et al., "Pathways and Modification of Warm Water Flowing beneath Thwaites Ice Shelf, West Antarctica," *Scientific Advances* 7, no. 15 (April 2021), https://doi.org/10.1126/sciadv.abd7254. The paper combines many of the data sets gathered onboard and proposes new pathways for warm water to be working beneath Thwaites.

211 ***The first time someone tossed*** Lars is referring to this study: Stanley S. Jacobs et al., "Antarctic Ice Sheet Melting in the Southeast Pacific," *Geophysical Research Letters* 23, no. 9 (May 1996): 957–60, https://doi.org/10.1029/96gl00723.

211 ***1.2 quintillion kilojoules of energy*** This is from a shipboard presentation of Anna's, "The Southern Ocean, the West Antarctic Ice Sheet, and an Orange Submarine." When I asked her for the reference

behind this number, she explained, "There is no paper, this is basic physics. Calculate the heat energy contained in the ocean (temperature times volume times density times specific heat capacity), then do the same for air: you will see that the air temperature needs to be over 400 degrees to have the same energy." She also provided the actual equation, but it is too cumbersome to include here.

211 ***the temperature of CDW is not uniform*** See this important and accessible paper that Lars cowrote with the oceanographer Karen J. Heywood and others about warming in the Southern Ocean: "Between the Devil and the Deep Blue Sea: The Role of the Amundsen Sea Continental Shelf in Exchanges between Ocean and Ice Shelves," *Oceanography* 29, no. 4 (December 2016): 118–29, https://doi.org/10.5670/oceanog.2016.104.

NAMELESS BAY

213 ***Never before has a human being*** Ali published a fascinating "behind the paper" article in which he describes just how unique it was for us to be in this part of the world, but he tells the story from the perspective of an Antarctic scientist: Ali Graham et al., "Orange Submarine 'Rán' Explores the Sea Floor in Front of Antarctica's Thwaites Glacier," *Nature*, September 5, 2022, https://earthenvironmentcommunity.nature.com/posts/orange-submarine-ran-explores-the-sea-floor-in-front-of-antarctica-s-thwaites-glacier.

215 ***"The Deluge"*** You can watch the talk here: Baylor Fox-Kemper, "Provost's Faculty Lecture Series: The Deluge," Brown University, YouTube, October 17, 2019, video, 58:18, https://www.youtube.com/watch?v=59HEYqKsLLM&t=3s.

220 ***Because it's cold*** Bill Watterson, "Why Does Ice Float?" *Calvin and Hobbes*, June 17, 1995, on GoComics, June 20, 2015, https://www.gocomics.com/calvinandhobbes/2015/06/20.

221 ***The procedural science*** Jeff Goodell wrote a nice piece about the coring that happened on our cruise: "Journey to Antarctica: The Dark Art of Coring," *Rolling Stone*, March 12, 2019, https://www.rollingstone.com/politics/politics-features/journey-to-antarctica-coring-thwaites-glacier-sea-pig-806769/. Linda wrote extensively about THOR's coring efforts in this detailed blog entry: Linda Welzenbach and Becky Minzoni, "Hammertime: THOR's Cores," *THOR Blogs (blog)*, Thwaites Glacier Offshore Research (THOR), March 18, 2019, https://

thwaitesglacieroffshoreresearch.org/blogs/2019/3/18/hammer time-thors-cores. And Tasha wrote about it here: "Snow on Ice: Sea Pigs and Mud #10," ITGC, March 22, 2019, https://thwaites glacier.org/blog/snow-ice-sea-pigs-and-mud-10.

222 ***the mud impounded in a Megacore*** The age of the mud impounded in each of these cores will depend largely on where the mud was found on earth. An area with little sedimentary activity will likely have older mud, closer to the surface. All of the age approximations relayed are specific to the Southern Ocean and, in particular, the Amundsen Sea.

222 ***the Last Glacial Maximum*** See this simple explanation for how the cryosphere and sea level have evolved in tandem over the last twenty thousand years: "How Does Present Glacier Extent and Sea Level Compare to the Extent of Glaciers and Global Sea Level during the Last Glacial Maximum (LGM)?" United States Geological Survey (USGS), accessed September 8, 2022, https://www.usgs.gov/faqs/how-does-present-glacier-extent-and-sea-level-compare-extent-glaciers-and-global-sea-level.

223 ***It's like flipping through pages*** Becky actually said this a little later in the cruise, while she was analyzing a different Kasten core, but because it is such a useful metaphor I have inserted it here, with her permission. On this first morning with mud onboard, she was very tight-lipped and focused, as the scene reflects.

223 ***within a scaffolding*** Cruikshank, *Do Glaciers Listen?*, 50.

223 ***The stories told in the "books"*** The oldest mud Becky studies from Antarctica is twenty thousand years old. Cuneiform is considered the first written language in human history and it was developed roughly six point five thousand years ago.

223 ***the last hundred years or so*** This information comes from Becky's dissertation: Becky Lynn [Totten] Minzoni, "The Antarctic Peninsula's Response to Holocene Climate Variability: Controls on Glacial Stability and Implications for Future Change" (PhD diss., Rice University, 2015), https://hdl.handle.net/1911/88211. She is currently working on publishing it as a book. Basically, her work shows that each glacier on the Antarctic Peninsula responds in a particular way in the context of its setting (oceanography, orography, size and shape of drainage basin, precipitation), such that some glaciers retreat while others advance. But the rapid warming over the last hundred years, in essence, trumps all of these local controls, and suddenly 90 percent of the peninsula's glaciers are retreating simultaneously.

225 ***organisms called forams*** Forams are fascinating. To learn about why studying forams is important, check out this informational

guide produced by the *JOIDES Resolution*, another famous oceanic research vessel: Deep Earth Academy, "Why Study Forams?," accessed September 9, 2022, https://joidesresolution.org/wp-content/uploads/2013/06/Mohawk-Additional-Background-3.pdf. For a deep dive into the particular forams found in Circumpolar Deep Water and what they tell us about deglaciation in Antarctica, see Elaine M. Mawbey et al., "Mg/Ca-Temperature Calibration of Polar Benthic Foraminifera Species for Reconstruction of Bottom Water Temperatures on the Antarctic Shelf," *Geochimica et Cosmochimica Acta* 283 (August 2020): 54–66, https://doi.org/10.1016/j.gca.2020.05.027.

226 ***among the first to analyze sediment from the Southern Ocean*** See John B. Anderson et al., "Sedimentary Facies Associated with Antarctica's Floating Ice Masses," in John B. Anderson and Gail M. Ashley, eds., *Geological Society of America Special Papers: Glacial Marine Sedimentation, Paleoclimatic Significance* (Boulder, CO: Geological Society of America, 1991). Anderson also wrote *the* textbook on this subject, *Antarctic Marine Geology* (Cambridge, UK: Cambridge University Press, 1999).

226 ***They tossed cores overboard*** Both Becky and Val Stanley, the Antarctic core curator at Oregon State University, emphasized just how profoundly the questions we have asked about Antarctica have changed over the last couple decades. See also Robert M. McKay et al., "Cenozoic History of Antarctic Glaciation and Climate from Onshore and Offshore Studies," in Fabio Florindo et al., eds., *Antarctic Climate Evolution* (Cambridge, MA: Elsevier, 2022).

226 ***Ready or not*** Lauryn Hill, vocalist, "Ready or Not," track 3 on Fugees, *The Score*, Ruffhouse Records, 1996.

228 ***Emilio Marcos Palma*** Perhaps the best single resource on Palma is, not surprisingly, Chipman, *Women on the Ice*, 117–18.

229 ***the main motive for Antarctic work*** The material in this section relies upon one of the most readable yet detailed history of humans in Antarctica that I have encountered: Tom Griffiths, *Slicing the Silence: Voyaging to Antarctica* (Cambridge, MA: Harvard University Press, 2007).

229 ***Argentina purposely deployed*** See Ignacio Javier Cardone, "Shaping an Antarctic Identity in Argentina and Chile," *Defense Strategic Communications* 8 (Spring 2020): 53–88, https://doi.org/10.30966/2018.riga.8.2.

229 ***Chile's dictator, Augusto Pinochet, visited*** They even made a stamp to commemorate the trip, which you can read more about in Jack Child, *Miniature Messages: The Semiotics and Politics of Latin American Postage Stamps* (Durham, NC: Duke University, 2008), 154.

230 ***Natalia López*** I interviewed Natalia on August 11, 2019. She is a journalist, writer, and curator. You can check her out on Instagram at @natitalo.

236 ***line items on a budget*** Hope Jahren's narrative nonfiction book *Lab Girl* (New York: Alfred A. Knopf, 2016) offers an entertaining and enlightening introduction to the ins and outs of fundraising for lab-based science, among other things.

236 ***data gathered on the continent must be made*** See this excellent introduction to the power of this unique agreement: "The Antarctic Treaty Explained," British Antarctic Survey, accessed September 9, 2022, https://www.bas.ac.uk/about/antarctica/the-antarctic-treaty/the-antarctic-treaty-explained/.

236 ***ask to withdraw a sample*** Go to the website for Oregon State's Marine and Geology Repository (https://osu-mgr.org/). There is big orange button with the words "Request Samples" right there in the middle of the landing page.

238 ***without the support of an ice sheet*** For more information on ice shelf buttressing, see G. H. Gudmundsson, "Ice-Shelf Buttressing and the Stability of Marine Ice Sheets," *Cryosphere* 7, no. 2 (2013): 647–55, https://doi.org/10.5194/tc-7-647-2013.

238 ***without the support of the sediment*** See Rush, *Rising*.

UNDERNEATH

249 ***like mere impressionistic paintings*** If you want to look at an excellent infographic showing how this works, check out the visuals that accompany Carolyn Beeler's dispatch here: "This Submarine's Historic Tour under Thwaites Glacier Will Help Scientists Predict Sea Level Rise," *The World*, May 20, 2019, https://interactive.pri.org/2019/05/antarctica/submarine-glacier-tour.html.

250 ***why doesn't it say you're not allowed to*** **make** ***someone pregnant*** I haven't been able to unearth this document, that being said, what matters here, I think, is Anna's perception of how women are treated in Antarctica: as burdensome others who are responsible alone for maintaining their sexual health and well-being while serving their governments in an isolated environment that doesn't support them. In 2022, the NSF released a report suggesting that the *majority* of people who present as women experienced sexual harassment or assault while on the ice. See: National Science Foundation (NSF) Office of Polar Programs (OPP) United States Antarctic Program (USAP), "Sexual Assault/Harassment

Prevention and Response (SAHPR) FINAL REPORT," June 22, 2022.

253 ***I* am *the process*** I am deeply indebted to the work of Iris Marion Young, perhaps best known for her article, "Throwing Like a Girl: A Phenomenology of Feminine Body Comport, Motility, and Spatiality," which was republished in the collection *On Female Body Experience: "Throwing Like a Girl" and Other Essays* (New York: Oxford University Press, 2005).

254 ***beech-like fossils that Robert Falcon Scott attempted*** Recent scholarship carried out by Paul Kenrick at the Natural History Museum in London shows that members of the Terra Nova Expedition likely encountered fossilized leaves of the Southern Beech on the glacial moraine of the Beardmore Glacier (as evidenced by the detailed notes of Edward Wilson, the expedition's famed chief scientist). However, no beech-like leaves were found among the thirty-five pounds of geologic samples discovered with the bodies of the deceased explorers. Subsequent expeditions have shown the rock in which these specific fossils are located is extremely delicate, and thus likely these specific fossils fragmented into unrecognizable pieces during Scott's ill-fated attempt to return from the pole.

254 ***his unwillingness to ditch*** Robert Falcon Scott died on his return from the South Pole, just eleven miles from One Ton Depot, his resupply camp. Recently some historians have hypothesized that he died because he was slowed down by the heavy fossils he carried with him and refused to jettison. For a deep dive into this controversial take, see Roland Huntford's *The Last Place on Earth: Scott and Amundsen's Race to the South Pole* (New York: Modern Library, 1999).

256 ***first detailed aerial footage ever shot*** I did, in fact, post this video to Twitter. You can see it here: Elizabeth Rush (@ElizabethaRush), "Today Carl Robinson of the British Antarctic Survey shared with us this FABULOUS (& unsettling) aerial footage of @ThwaitesGlacier," Twitter, February 21, 2019, 6:34 p.m., https://twitter.com/ElizabethaRush/status/1098742857874243584.

260 ***collecting all kinds*** If you want to read the first paper published as a result of the data collected on this day, see Wåhlin et al., "Pathways and Modification."

265 ***Ali sits beside Jonas*** Tasha wrote about this in her blog: "Snow on Ice: Synergizing Science #9," ITGC, March 15, 2019, https://thwaitesglacier.org/blog/snow-ice-synergizing-science-9.

265 ***They're corrugation ridges*** Ali would go on to publish a really important paper using these ridges as proof that sometime over the past couple centuries, Thwaites deteriorated at rates two to three times faster than we see today: Alastair G. C. Graham et al., "Rapid Retreat of Thwaites Glacier in the Pre-Satellite Era," *Nature Geoscience* 15 (2022), 706–13, https://doi.org/10.1038/s41561-022-01019-9.

272 ***In premodern times*** Randi Hutter Epstein, *Get Me Out: A History of Childbirth from the Garden of Eden to the Sperm Bank* (New York: W. W. Norton, 2010).

272 ***makers of death*** See Claudia Dey, "Mothers as Makers of Death," *Paris Review*, August 14, 2018, https://www.theparisreview.org/blog/2018/08/14/mothers-as-makers-of-death/. In this essay, Dey reflects on the process of writing her novel *Heartbreaker* during an intense ten-day spell, during which she was recovering from ear surgery and her two young children were on a road trip with their father. It is an essay that explores the contours of holding onto a writing life while also caring for small human beings and the pitfalls of always thinking about motherhood as something innocent and pure. Her thinking in that piece certainly informs my own here.

THE QUICKENING

282 ***Cohesive shelf. Exploded lodestar*** Linda posted these and other images on her blog: Linda Welzenbach, "Thwaites Glacier: There and Gone," *THOR Blogs* (blog), Thwaites Glacier Offshore Research (THOR), March 12, 2019, https://thwaitesglacieroffshoreresearch.org/blogs/2019/3/12/thwaites-glacier-there-and-gone.

282 ***the Larsen B collapse*** After scientists identified West Antarctica as a potential source of significant sea level rise in the early eighties, there was a debate about whether the pace of ice loss was sufficient to fundamentally change human civilization in the coming centuries. Some, like Charles Bentley, the first person to map the depth and extent of Thwaites, said no. That the disintegration of an ice shelf would not accelerate the rate at which surrounding glaciers made their way into the sea. But he would be proven wrong by what happened as Larsen B collapsed, and all the glaciers in the region sped up. See "Assessing the Ice," in Oppenheimer et al., *Discerning Experts: The Practices of Scientific Assessment for*

Environmental Policy (Chicago: University of Chicago, 2019). And for incredible images of the Larsen B collapse, see "World of Change: Collapse of the Larsen-B Ice Shelf," NASA Earth Observatory, accessed September 20, 2022, https://earthobservatory.nasa.gov/world-of-change/LarsenB.

287 ***To let the baby out*** Maggie Nelson, *The Argonauts* (Minneapolis, MN: Graywolf Press, 2015), 124.

288 ***Motherhood is the adventure now*** When I embarked on this adventure, I reached out to a phenomenally wise midwife friend, Miriam Rosenburg, and she suggested some reading. Here is an abridged list of the books that I found the most engaging and helpful: Emily Oster's *Expecting Better: Why the Conventional Pregnancy Wisdom Is Wrong—and What You Really Need to Know* (New York: Penguin, 2013), Anne Lamott's *Operating Instructions: A Journal of My Son's First Year* (New York: Anchor Books, 2005), Peggy Vincent's *Baby Catcher: Chronicles of a Modern Midwife* (New York: Scribner, 2002), Angela Garbes's *Like a Mother: A Feminist Journey through the Science and Culture of Pregnancy* (New York: HarperWave, 2018), and Janet Schwegel's *Adventures in Natural Childbirth: Tales from Women on the Joys, Fears, Pleasures, and Pains of Giving Birth Naturally* (New York: Marlowe and Company, 2005).

289 ***Why is no woman's labor*** Louise Erdrich, *The Blue Jay's Dance: A Memoir of Early Motherhood* (New York: Harper Perennial, 2010), 35.

297 ***It comes from the Old English*** See *Online Etymology Dictionary*, s.v. "quicken (*v.*)," accessed September 10, 2022, https://www.etymonline.com/word/quicken.

HOLDING SEASON

303 ***a single confirmed case of COVID-19*** There are lots of news sources I could reference to draw a map of what was happening in the early pandemic, but I think Ann Patchett gets at the experience of what it was like for someone like me—that is, someone with a secure job that allowed them to work from home—in her essay "These Precious Days," *Harper's*, January 2021, https://harpers.org/archive/2021/01/these-precious-days-ann-patchett-psilocybin-tom-hanks-sooki-raphael/.

307 ***Who named the leopard seal*** The actual story of how the leopard seal got its name is not clear; it may have been thanks to a globe-trotting big game hunter but then again it very well might not.

310 ***It is the same story*** There is a lot of great writing about the intersection of climate change and racial injustice, some of which touches on COVID. Writers who have been particularly influential for me are Emily Raboteau and Mary Annaïse Heglar, both of whom have work included in this excellent roundup: "Summer Reading List on Climate Change," Columbia Law School, accessed September 10, 2022, https://climate.law.columbia.edu/content/summer-reading-list-climate-justice.

310 ***In the absence of any reliable centralized response*** See this powerful essay on COVID, race, and community: Mary Louise Pratt, "Airways," *Contactos*, accessed September 10, 2022, https://contactos.tome.press/airways/.

316 ***the scientists were able to complete*** If you are interested in reading the 181-page cruise report, which offers a detailed account of all we accomplished, see R. D. Larter et al., "Cruise Report: RV/IB *Nathaniel B. Palmer*, NBP 19-02," 2019, https://service.rvdata.us/data/cruise/NBP1902/doc/NBP1902_report_final.pdf?f=cruise/NBP1902/doc/NBP1902_report_final.pdf.

GOING TO PIECES

331 ***the organization where I volunteer*** I volunteered with Nationalize Grid, a branch of the Providence chapter of the Democratic Socialists of America, though during the pandemic and early motherhood my participation declined precipitously. On the issue of utilities shutoffs, as on the issue of who would hold the position of public utilities chair in Rhode Island, we partnered with the George Wiley Center in Pawtucket, an organization that has been advocating for social justice in the region since 1981.

331 ***a moratorium on utilities shutoffs*** See the bill that passed through the Rhode Island house of representatives here: "An Act Relating to Public Utilities and Carriers—Termination of Service during Periods of Declared Emergencies," H.R. 5442 (2021), http://webserver.rilegislature.gov/BillText21/HouseText21/H5442.html.

343 ***the nearby Gregorio refinery*** Fossil fuel extraction, sheep farming, and tourism are the three main economic drivers in Patagonia. To learn more about the impact fracking is having on the region, see Nina Negron, "Fracking Leaves Heavy Footprint in Argentina's Patagonia," Phys.org, December 13, 2019, https://phys.org/news/2019-12-fracking-heavy-footprint-argentina-patagonia.html.

344 ***if I wish a child into this world*** This line is inspired by a line ("I wished you into / this world, so / I cannot say / I didn't wish / this world upon you") in Ambalila Hemsell's poem "Passport," in *Queen in Blue* (Madison: University of Wisconsin Press, 2020). This breathtaking collection is one of the best meditations on motherhood and carrying a child into a damaged world where our complicity grows deeper daily.

EPILOGUE

348 ***Take ANDRILL here*** The Exploratorium in San Francisco hosts a pretty cool video on the ANDRILL project on their website: "The ANDRILL Project," Exploratorium, March 15, 2006, video, 38:53, https://www.exploratorium.edu/video/andrill-project.

352 ***the invasive woolly adelgid*** For more on woolly adelgid, hemlock trees, and climate change, see Annie Paradis et al., "Role of Winter Temperature and Climate Change on the Survival and Future Range Expansion of the Hemlock Woolly Adelgid (*Adelges tsugae*) in Eastern North America," *Mitigation and Adaptation Strategies for Global Change* 13, no. 5 (June 2008): 541–54, https://doi.org/10.1007/s11027-007-9127-0.

352 ***a broad-leafed black oak*** Black oaks appear to be one of the northern trees most suited to changes in the climate. See James Theuri, "Will Black Oaks Survive Climate Change in the Midwest?," Trees for Energy Conservation, September 10, 2019, https://trees-energy-conservation.extension.org/will-black-oaks-survive-climate-change-in-the-midwest/.

352 ***a recently arrived red-bellied woodpecker*** I read about the expansion of the red-bellied woodpecker's range in one of my local newspapers: Todd McLeish, "How Climate Change Is Impacting Bird Populations in Newport County," *Newport Daily News*, January 8, 2021, https://www.newportri.com/story/lifestyle/2021/01/08/climate-change-impacts-bird-populations-newport-county-rhode-island/4132447001/.

352 ***the number of scarlet tanagers*** The Audubon Society has been studying climate change's impact on birds for years. Theirs is one of the most extensive records we have. Check out their recent study, "Survival by Degrees: 389 Bird Species on the Brink," accessed September 10, 2022, https://www.audubon.org/climate/survivalbydegrees.

352 ***Deep, interconnected troughs*** Kelly A. Hogan et al., "Revealing the Former Bed of Thwaites Glacier Using Sea-Floor Bathymetry:

Implications for Warm-Water Routing and Bed Controls on Ice Flow and Buttressing," *Cryosphere* 14, no. 9 (2020): 2883–908, https://doi.org/10.5194/tc-14-2883-2020.

352 ***a clockwise circulation*** Wåhlin et al., "Pathways and Modification."

353 ***Aerial evidence of growing cracks*** This information comes from an early synopsis of some of the fieldwork being done at Thwaites, titled "Collapse of Thwaites Eastern Ice Shelf by Intersecting Fractures," during a press conference at the 2021 meeting of the American Geophysical Union. Read a write-up about it here: Paul Voosen, "Ice Shelf Holding Back Keystone Antarctic Glacier within Years of Failure," *Science*, December 13, 2021, https://www.science.org/content/article/ice-shelf-holding-back-keystone-antarctic-glacier-within-years-failure.

353 ***the ice shelf didn't regenerate*** See Braddock et al., "Relative Sea-Level Data Preclude Major Late Holocene Ice-Mass Change in Pine Island Bay." One of the highlights of my career is having my field-work assistance acknowledged in this paper. Those couple of days I spent on the islands with Scott and his team are burnt into my heart and mind in the best way possible. Thank you for letting me help.

353 ***corrugation ridges*** Graham et al., "Seabed Corrugations beneath an Antarctic Ice Shelf."

353 ***complete comprehension wouldn't be possible*** In his talk "The Deluge," Baylor Fox-Kemper talks about the complexity of our planetary system and how hard it is to pack all the details we know, let alone those that we are just now discovering and recording, into even our most comprehensive climate models.

353 ***we thought would happen in the future*** As climate change has evolved over recent decades, increased signs of anthropogenic forcing have become stronger and stronger, which has resulted in a wide range of research. The IPCC recently published what amounts to a literature review of how our understanding of this phenomenon and the science we use to better understand it has changed over time: Hervé Le Treut et al., "Historical Overview of Climate Change Science," in S. D. Solomon et al., *AR4 Climate Change 2007: The Physical Science Basis* (Cambridge, UK: Cambridge University Press, 2007), https://www.ipcc.ch/site/assets/uploads/2018/03/ar4-wg1-chapter1.pdf.

353 ***outdated before they even get underway*** One of the largest challenges with climate change adaptation is how to design for a world that is changing at such a surprising rate. One way to do so is to be transparent about what isn't working as well as what is. See Ross Westoby et al., "Sharing Adaptation Failure to Improve

Adaptation Outcomes," *One Earth* 3, no. 4 (2020): 388–91, https://doi.org/10.1016/j.oneear.2020.09.002.

353 ***sweet, almost harmless*** *The Climate Pod* interviewed Richard Alley, one of the world's leading glaciologists and an expert on Thwaites. He spoke at length about how, given all we don't know about the glacier, we could be significantly underestimating the worst-case scenario. Check out the episode: Brock Benefiel and Ty Benefiel, "The 'Doomsday Glacier's' Disastrous Potential (w/ Dr. Richard Alley)," January 5, 2022, in *The Climate Pod*, podcast, 53:58, https://www.theclimatepod.com/episodes/episode/2a1fcfd4/the-doomsday-glaciers-disastrous-potential-with-dr-richard-alley.

354 ***As modelers fit*** This information comes from a phone call I had with Jeremy Bassis in spring 2021, as well as from an article he wrote, "Quit Worrying about Uncertainty in Sea Level Projections," *Eos* 102 (November 2021): 19–21, https://eos.org/opinions/quit-worrying-about-uncertainty-in-sea-level-projections.

354 ***flexible framework for just climate*** How exactly we ought to do this is the trillion-dollar question. Allocating money for adaptation isn't enough; those living on climate change's front lines must maintain and even gain agency as they evolve in step with external stressors. Anything less results in a less equitable world. Here are a couple of organizations already doing this important work: Anthropocene Alliance (https://anthropocenealliance.org/), which connects frontline climate changed communities to each other and to the resources to help them fight for a just recovery; LA Safe (https://lasafe.la.gov/) in Louisiana, which has designed and funded community-centered sea level–rise response efforts, and Adapt Alaska (https://adaptalaska.org/), which focuses on Indigenous-centered adaptation outreach and implementation.

354 ***I have kept track*** The framing of this final section is inspired by the last section of Joan Didion's showstopper essay, "The White Album," in *The White Album* (New York: Simon & Schuster, 1979). It is widely anthologized, and I first encountered it in John D'Agata, ed., *The Next American Essay* (Minneapolis, MN: Graywolf Press, 2003).

355 ***4,264 vertical profiles*** This information comes from Lars Boehme and Isabella Rosso, "Classifying Oceanic Structures in the Amundsen Sea, Antarctica," *Geophysical Research Letters* 48, no. 5 (March 2021), https://doi.org/10.1029/2020GL089412. Just for comparison, during the month and a bit that we were in the region, we completed 104 CTD casts, or roughly one-fortieth of the data that the seals contributed.

Acknowledgments | *Antarctic Thanks*

All creative acts are communal. Nothing is crafted alone. To Felipe Martínez Pinzón—*mi socio* in all things from making a life and making a home to making the time to make a book and the hundreds of conversations that fed it—you are my chief collaborator. None of this would be possible without you. To you: sweetness, solace; light and lighthouse, illuming these precious days.

To my shipmates: this book literally would not be but for the generous gift of your time, attention, expertise, and care. When you signed up to travel to the bottom of the planet on an icebreaker bound for Thwaites, you had no idea what you were getting into, not really. Thank you for sitting down with me, again and again and again—during what were likely some of the most important months of your life—to share your thoughts on glacial collapse, regeneration, and the dinner menu. Thank you for reviewing this manuscript multiple times during the years since our return. The poetry in these pages is yours.

To my family, including the one I was gifted when Felipe and I married, thank you. Mom and Dad: your love—the endlessness of it—carried me all the way to the calving edge of Antarctica. Aunt and Unc: I am so lucky to count you as my second parents. To Maria Elisa and Ernesto: thank you for opening your home in the middle of the pandemic. I am forever grateful that we were able to spend those otherwise isolating months together. To Rosario: thank you for the fashion advice and also for raising such an astonishing man. To my godmother, Leah Brecher: thank you for your radiance and typo-targeting attention to detail. To Nico, the elder: I couldn't ask for a better brother-in-law. To Nico, the younger: this glittering world is even brighter with you in it. Thank you for choosing to spend your earthly time with us.

This book was largely written and revised during the Pandemic, during which time a few special people cared for my son while I stepped away to write. To Elvira Villamil, Patty Rodríguez, Milena Vargas, my parents and my mother-in-law: you have given me something so few new mothers get, peace of mind. Nico has spent the first two years of his life held—in body, in mind, and in heart—by you.

To Elise Bonner: when you moved to Providence, my adult life suddenly made more sense. Thank you for the walks, the talks, the at home yoga classes, and the ladies burger and beer nights. I adore growing alongside you. To Christopher Gupta: thank you for all the good cooking, including a parmesan loaf a week which kept this very-pregnant-in-the-middle-of-a-pandemic-lady very happy and nourished. To Sarah Thomas and Dani Blanco, for simultaneously being reliable and really fun: thank you for the fabulous company and for watering the plants. To Lauren Lanahan: my sister from another mother, your presence on this planet makes me feel less alone.

While writing this book I became a pregnant person, then a laboring person, then a parent. I'm not big on transitions and I am pretty sure I would not have survived these if it weren't for the thoughtful guidance and care I received from a whole crew of wise people. To Miriam Jagle: dear friend and midwife extraordinaire, not only did you read an early draft of this book, you sent me into the world with a killer motherhood reading list *and* patiently answered my questions about eating blue cheese and early trimester spotting, about home births and breastfeeding. To Kaeli Sutton: your birth classes were life changing. I only wish that everyone—pregnant and otherwise—got the chance to prepare for big changes in such a thoughtful manner. To my birth doula, Lisa Gendron: your quiet advice and unflagging confidence helped me make it through labor. To Dr. Morton: when the world was going to pieces, you made me laugh, again and again, and then you stood back as I delivered my baby. Thank you for your levity and the space.

To Fred Swanson at the H.J. Andrews Experimental Forest: thank you for inviting me to dwell among the towering conifers in Oregon's central cascades where I walked and wrote my way through an early draft of this book. Thank you to Lisa Sackreiter for bringing soup and salad to my doorstep so I could do a single day mini-residency in the murky middle. Thank you to Megan Mayhew Bergman for inviting me to teach at Bread Loaf in Vermont where glossy green peaks kept me company in the early morning as I more-or-less finished this book.

To Joey McGarvey: How do I even begin to thank you? When I thought about how difficult it would be to write a book and bring a baby into the world at the same time, I knew that you would be the best literary traveling companion during this demanding journey. How incredibly grounding it was to have an editor who understood the ambition of this project while still posing really difficult questions about what it could become; an editor who helped to envision the big picture then, years later, mussed over minutia until sentence after sentence sung. My writing, my life, would be so much less without you.

To Milkweed Editions: you all are my publishing family. Yanna Demkiewicz, thank you for promoting Milkweed books with an unparalleled amount of panache. To Daniel Slager: thank you for your support and for teaching me about book publishing. Bailey Hutchinson, it has been a pleasure partnering with you through the production process. Quinton Singer: that I can sleep at night has everything to do with your thorough fact-checking. To Mary Austin Speaker: your book designs are art in their own right. Your patience and creativity made this strange form make sense on the page.

To the fabulous extended network that supports all this good labor: to Julia Lord, for believing in me, in this book, from the start. To Christie Hinrichs of Author's Unbound, you are a godsend. My inbox, my travel, my life has become easier with you at the helm of all things away. To Whitney Peeling, for casting the net wide and for inviting as many readers as possible in, thank you for championing this book. And to you, dear readers, without whom none of this would have been possible.

One of the greatest gifts the writing life has yielded is that of knowing other writers. To Meera Subramanian, Kerri Arsenault, Bathsheba Demuth, and Sumanth Prabhaker: my early readers and dear friends. Each of you engaged with some portion of this project (or all of it, I see you Meera) at some pivotal point and provided much-needed feedback. To my blessed writing group—Alizah Holstein, Nate McNamara, Jodie Noel Vinson, Sarah Frye—you read bits of what would become this book in their earliest manifestations and asked the questions that would guide the whole drafting process.

As I wrote, the work of many other mother-writers kept me company, opened me up, and reminded me of what is possible: Jazmina Barrera, Belle Boggs, Eleanor Davis, Camille T. Dungy, Louise Erdrich, Angela Garbes, Ursula Le Guin, Alexis Pauline Gumbs, Ambalila Hemsell, Anne Lamott, Audre Lorde, Lorrie Moore, Megan Mayhew Bergman, Meaghan O'Connell, Emily Raboteau, Sarah Ruhl, Peggy Vincent, Rachel Zucker, and so many more who went before and paved the way for what has come. I am also deeply grateful to Krys Malcolm Belc for writing *The Natural Mother of the Child: A Memoir of Nonbinary Parenthood.*

Portions of this book appeared elsewhere and in other incarnations. "Searching for Women's Voices in the Harshest Landscape on Earth" was published in *LitHub* and was later anthologized in *The Earth as We Knew It;* "First Passage" found a home at *Orion Magazine;* "What does Antarctica's Disintegration Ask of Us" was published in the *New York Times;* and something about time and shifting landscapes will soon appear in *Emergence Magazine.* Thank you to all the editors with whom I have collaborated over the years.

Thank you to the various organizations who supported this work: National Geographic, the Alfred P. Sloan Foundation, Kari

Traa, Zoom audio recorders, the National Science Foundation, and in particular Valentine Kass.

To my students: for working so hard and opening up; for your labor, trust, and time. It's thanks to you that my relationship to writing remains a process of becoming. Thank you in particular to Liza Yeager, Claire Boyle, and Ivy Scott, you three, continue to teach me so much.

To the oceans, ice, and rock at the bottom of the planet: thank you for giving us each other, thank you for teaching me something about what it means to be held—rapt, in cold suspension, wholly. This book is an offering to you.

photo: Stephanie Ewens

ELIZABETH RUSH is the author of *Rising: Dispatches from the New American Shore*, which was a finalist for the Pulitzer Prize, and *Still Lifes from a Vanishing City: Essays and Photographs from Yangon, Myanmar.* Her work has appeared in the *New York Times*, *Harpers*, *Orion*, *Granta*, *Guernica,* and others. A recipient of fellowships from the National Science Foundation, National Geographic, the Andrew Mellon Foundation, the Howard Foundation, and the Metcalf Institute, Rush lives in Rhode Island with her husband and son, where she teaches at Brown University.

Founded as a nonprofit organization in 1980, Milkweed Editions is an independent publisher. Our mission is to identify, nurture, and publish transformative literature and build an engaged community around it.

Milkweed Editions is based in Bdé Óta Othúŋwe (Minneapolis) within Mní Sota Makhóčhe, the traditional homeland of the Dakhóta people. Residing here since time immemorial, Dakhóta people still call Mní Sota Makhóčhe home, with four federally recognized Dakhóta nations and many more Dakhóta people residing in what is now the state of Minnesota. Due to continued legacies of colonization, genocide, and forced removal, generations of Dakhóta people remain disenfranchised from their traditional homeland. Presently, Mní Sota Makhóčhe has become a refuge and home for many Indigenous nations and peoples, including seven federally recognized Ojibwe nations. We humbly encourage our readers to reflect upon the historical legacies held in the lands they occupy.

milkweed.org

Milkweed Editions, an independent nonprofit publisher, gratefully acknowledges sustaining support from our Board of Directors; the Alan B. Slifka Foundation and its president, Riva Ariella Ritvo-Slifka; the Amazon Literary Partnership; the Ballard Spahr Foundation; *Copper Nickel*; the McKnight Foundation; the National Endowment for the Arts; the National Poetry Series; and other generous contributions from foundations, corporations, and individuals. Also, this activity is made possible by the voters of Minnesota through a Minnesota State Arts Board Operating Support grant, thanks to a legislative appropriation from the arts and cultural heritage fund. For a full listing of Milkweed Editions supporters, please visit milkweed.org.

Interior design by Mary Austin Speakers and Tijqua Daiker
Typeset in Caslon

Adobe Caslon Pro was created by Carol Twombly
for Adobe Systems in 1990. Her design was inspired by
the family of typefaces cut by the celebrated engraver
William Caslon I, whose family foundry served
England with clean, elegant type from the early
Enlightenment through the turn of the
twentieth century.